THE BRAIN'S BODY

THE BRAIN'S BODY

Neuroscience and Corporeal Politics

VICTORIA PITTS-TAYLOR

DUKE UNIVERSITY PRESS

Durham and London

2016

Typeset in Minion Pro by Copperline

Library of Congress Cataloging-in-Publication Data
Pitts-Taylor, Victoria, author.
The brain's body : neuroscience and corporeal politics /
Victoria Pitts-Taylor.
pages cm
Includes bibliographical references and index.
ISBN 978-0-8223-6107-7 (hardcover)
ISBN 978-0-8223-6126-8 (pbk.)
ISBN 978-0-8223-7437-4 (e-book)
1. Neurosciences—Social aspects. 2. Cognitive neuroscience—
Social aspects. 3. Feminist theory. I. Title.
RC343.3.P58 2016
612.8'233—dc23
2015033351

Cover art: Scott Camazine. Fiber tractography
image of the human brain. © Scott Camazine /
Alamy Stock Photo, detail.

FOR EMERSON TAYLOR AND CHLOE TAYLOR

CONTENTS

ACKNOWLEDGMENTS

I thank many people and organizations for their assistance with this project. The journal *Hypatia* published an earlier version of chapter 3, as "I Feel Your Pain: Embodied Knowledges and Situated Neurons" (2013). I thank the journal and its editors and reviewers for feedback. My editor at Duke University Press, Courtney Berger, offered invaluable guidance, as did three anonymous readers. A Mellon Foundation midcareer faculty fellowship supported my participation in a seminar run by the Committee for Interdisciplinary Science Studies at the Graduate Center, City University of New York, whose members gave feedback on several chapters of this book. So did the members of the NeuroCultures Seminar that ran from 2010–12 at the Graduate Center, which Jason Tougaw and Rachel Liebert coorganized with me. Members of the Neurogenderings Network, whose conferences I attended in Vienna and Lausanne, offered insight into the possibilities for feminist neuroscience and positive critique. The Center for the Study of Neuroscience and Society at the University of Pennsylvania provided funding for me to attend their residential training for nonscientists in 2010. I also thank CUNY and the Department of Sociology at Queens College for funding to attend the University of Pennsylvania. The Hastings Center for Bioethics provided a beautiful space to work during my sabbatical. I am grateful for helpful conversations and feedback from colleagues and friends, including Alexandre Baril, Maria Brincker, Jeffrey Bussolini, Patricia Clough, Christina Crosby, Lennard Davis, Ashley Dawson, Peter Godfrey-Smith, Lori Gruen, Siri Hustvedt, Rebecca Jordan-Young, Eben Kirksy, Emily Martin, Martin Pickersgill, John Protevi, Rayna Rapp, Joe Rouse, Deboleena Roy, Laura Sizer, Jason Tougaw, Bryan Turner, Kari Weil, and Elizabeth Wilson. I thank Jesse Prinz for reading several of the chapters and giving nonstop encouragement. For their sustaining friendship I thank Joe Rollins, Elizabeth Wood, and Amy Krakowka. For their constant support I thank my sisters Jennifer Gosetti-Ferencei and Angela Pitts. For practical help along the way, I am grateful to Marietta and Roger Beougher, Jerry Pitts, Janet Taylor McCracken, Barbara and Mark Taylor, JoAnne and Kenneth Taylor, and Katie Larimer. I offer special thanks to my partner Gregory Taylor, whose dual perspective as a clinician and scientist allowed us to have rich conversations about matters discussed here. Finally, I thank Chloe Taylor and Emerson Taylor. They taught me the paradox of plasticity: the remarkable openness of possibility and the entanglements in time, space, and experience. This book is for them.

Introduction

The Social Brain and Corporeal Politics

PRELUDE

The brain matters. How it matters in social thought is one concern of this book. How it literally matters, or becomes itself, is another. These questions of knowing and being, and of discourse and biology, are interrelated. We need to understand "how meanings and bodies get made," as Donna Haraway wrote, "not to deny meanings and bodies, but in order to build meanings and bodies that have a chance for life" (1988, 580). This book grasps meanings and bodies together. The meanings include ideas about nature and culture, universality and difference, normalcy and pathology, plasticity and transformation, sociality and kinship, power and inequality, to name a few. The bodies are bodies with brains: neurobiological bodies, bodies conceived in relation to the central nervous system, to neurons, neurotransmitters, and brain regions. They are also minded bodies, whose psychic capacities are material and physical as well as phenomenological and situated. They are social bodies, too, which get made not only in evolutionary time but also in the immediate past and present, in relation to the other bodies (and things) with which they coexist. The task is to understand their making and meaning, without denying them realness. The task is also to pay attention to what "chances for life" are at stake in our attempts to understand and manage them.

The Social Brain

The brain is now a social problem in at least two senses. The brain is conceived as the biological ground for the self and social life, as both constituting the mind and underpinning intersubjectivity. This has encouraged many neuroscientists in recent years to speak as philosophers and sociologists. The brain is also understood as *itself* a product of sociality, built through experience and open to transformation.

The prevailing scientific conception of the brain, as everyone knows, is plastic. *Plasticity* refers to the brain's ability to biologically change and be changed, not merely on a phylogenetic scale but also an ontogenetic scale. It means that the brain is not hardwired, but rather constantly developing and changing in response to experience. The plastic brain can be understood as nurtured as well as natured, and thus as a mode for and a reflection of environmental influences on the body and self. It is now possible to argue, "Each person's brain has a great degree of plasticity and develops uniquely in response to the social and natural environment as the person develops over the course of his or her life" (Glannon 2002, 250). This means, in principle, that neurobiology is not universal and predictable, but rather differentiated through experience. Contemporary depictions of the "social brain" also see it as relational, "in dynamic interaction with the environment" (250). Advocates of social brain research argue that while humans are "constituted by evolutionary biology," they also must be seen as "embedded in complex social networks," "largely habitual," "highly sensitive to social and cultural norms," and "more rationalising than rational."[1]

The plastic, social brain challenges the separation of mind and body, culture and nature, that is characteristic of twentieth-century thought. Throughout much of the past century, in many different fields of inquiry, the mind was treated independently of the brain. The mind once was understood in cognitive science, for example, as a problem of abstract computation that could be modeled by computers, without attending to the capacities of a fleshly, biological organ. In the humanities and social sciences the mind often was addressed through rationalism or psychological drives, or in terms of symbolic interaction, cultural inheritance, socialization, or discourse and subjectivation. These different perspectives assumed the brain to be fixed hardware—necessary for, but inessential to the study of, cognition and culture. The brain belonged to the biological body, whereas the mind was understood as immaterial, symbolic, intellectualist, or discursive. A focus on symbolic representation allowed studies of consciousness, the psyche, the self, the subject, and behavior to proceed without reference to the brain or neurophysiology.[2] Crucially, this also enabled a more or less strict division between biology, as that which is given by nature, and politics, as that which is amenable to social contest and transformation.

All of these assumptions are being challenged. The mind is now widely defined as equivalent to, or as emerging from, the brain. This neuroscientific turn began with biological psychiatry, which transformed the view of mental illness into the study of neurochemical disorder. Then, in the 1980s, a new model of cognition, called connectionism, allowed cognitive science to become relevant to the study of the organic brain. Over the next two decades, the field's boundaries with neuroscience eroded. The brain sciences have since made deep inroads into philosophy and other disciplines interested in the mind. Emboldened by paradigm shifts, new experimental methods, and the advent of functional brain imaging technologies, the brain sciences also have spawned subfields such as social neuroscience and cultural neuroscience. These aim to study not only the individual mind/brain but also intersubjectivity, social relations, and cultural differences through neuroscientific methods. Massive financial investments are buttressing these efforts to explore human experience via neurons, neurotransmitters, neural networks, and brain regions.[3]

The brain sciences also inform contemporary understandings of the self. Nikolas Rose and Joelle Abi-Rached (2013) argue that in the twenty-first century, neuroscientific and neurobiological concepts play a prominent role in identity. Rose (2006) depicts the contemporary "neuromolecular" style of thought as an extension of the somaticization of the individual that began in the late twentieth century. Somaticization brings the physical body and health to bear on conceptions of the self, personal identity, and citizenship.[4] Neuroscience is opening up that management of the self to new brain-based practices of care, enhancement, and optimization, including pharmaceutical interventions. But far from individualizing the human and reducing her to neurons or brain chemistry, the neurosciences are embracing a fundamentally social conception of the brain. Thus, Rose and Abi-Rached argue, the ascendance of neuroscience is not "a fundamental threat to the human and social sciences," but rather an "opportunity for collaboration across the two cultures" (142).[5]

These reassurances notwithstanding, the prospect of neuroreductionism remains deeply worrying for many critics, as I discuss below. But while the mind has been materialized, understood as equivalent to or emerging from the brain, the embrained subject is not fully "decapitated" (E. Wilson 2004). In neurocognitive science and philosophy of mind, for example, theories of embodied, situated cognition see the brain/mind in relation to the body as a whole and as modified by its environment. Neurocognitive theories of emotion argue that the body is a mediator and repository of memory and emotional valence. Through "somatic markings" (Damasio 1994, 1996) and "embodied appraisals" (Prinz 2004), the sensory and feeling body participates in the generation of meaningful experience. In the theory of extended cognition (Clark 2001, 2008b), grounded in research on brain–machine interfaces and prosthetics, the brain adapts and transforms through pragmatic bodily activity in the environment, leading to multiple brain/body/world assemblages. In the neuroscientific research on mirror neurons the brain is depicted as inherently social, crossing the boundary of the individual through neural processes that simulate the other, as well as through motor schema that some propose to be collectively shared by conspecifics.[6] These accounts have been hugely influential among social theorists because, to varying degrees, they are seen to challenge biological determinism and reductionism while accounting for the bodily materiality of mind and experience.

The focus on embodiment also enables an ethical appeal, potentially addressing issues of human suffering or emphasizing the mutual interdependence and vulnerability of persons. For example, "a biological perspective on social relations may help us to relate to people as complex living systems influenced by other systems, rather than as disembodied Cartesian 'minds' cut off from the world. Perhaps this perspective gives a firmer empathetic basis to connect with each other through our shared embodiment, with all that this means in terms of susceptibility to pain, hunger, stress, loss, mortality, illness and joy" (Rowson 2011). By attending to the physical, neurobiological body, the argument goes, we can grasp a richer sense of persons, their interconnectedness with others, and their embeddedness in the environment. As I discuss later, the literature on embodiment in neuroscience and naturalized philosophy even resonates to some degree with feminist epistemologies, which argue for the significance of emotions and the "situatedness" of the mind in the body and world. (There are also, I will argue, instructive differences between and within these literatures.)

The social brain, in sum, is equated with mind and self, is understood as the product of embodied experience, is seen to provide the foundation for (and is reflected in) social structures, and is subject to intervention and transformation. In these ways the social brain collapses the distinction between nature and culture. Although it is not framed as such in scientific accounts, the plastic, social brain also reveals neurobiology to be political— that is, capable of change and transformation, and open to social struc- tures and their contestation.[7] This book is concerned with the corporeal politics of the brain and the neurobiological body. I address, on the one hand, how social norms, power, and inequality affect representations of the brain and, on the other, how they are understood to literally entangle with neurobiological processes. I draw from feminism, queer theory, disability studies, feminist science studies, philosophy, and sociology, as well as the neurosciences, to address the limits of current scientific conceptions of the social brain, and also to explore the queerer possibilities that can be found in neurobiological data.[8]

Science and Critique

The centrality of the brain in contemporary thought and public life—the "neuroscientific turn"—is hugely worrying for many critics. To begin, it raises the dual specters of biological determinism and neuroreductionism. Biological psychiatry has long been faulted for treating mental illnesses exclusively in terms of receptors, pathways, and neurochemicals, while ig- noring or downplaying biographical and social contributions to mental suffering. The more fundamental reduction of all mental content to neural processes can have the effect of dismissing social and cultural aspects of the mind, psyche, and behavior. It may impose universalized conceptions of biology onto mind and cognition, and pathologize those who do not meet normative brain measures. Neuroreductionism, the view that "human be- ings are essentially their brains" (Vidal and Ortega 2011, 7), also marginal- izes fields that use hermeneutic and interpretive methods, rather than nar- rowly empirical ones, and has the potential to dominate the understanding of health, wellness, and personal identity in everyday life.[9] Further, it can enable the naturalization of social inequalities. Historically, scientific asser- tions about human nature have been "a politics by other means" (Asberg and Birke 2010, 414), used to pathologize women and racial and ethnic minorities, and the threat of "neurosexism" and other problems looms large (Jordan-Young and Rumiati 2012). The reduction of social problems to neurobiology can also result in the subjection of individuals and groups to techniques of neurogovernance.

One of the most resilient examples of both neuroreductionism and biological determinism is the sexed brain—that is, the brain that is sexed prenatally by reproductive hormones, is organized for reproduction, and shapes gender roles. Even in the era of neural plasticity, some sex difference researchers claim not only that the brain is organized in utero as male or female but also that this organization shapes individuals' gender identity, sexual orientation, and cognitive traits (e.g., Bao and Swaab 2011). Although they could be read otherwise, studies using brain imaging technologies to detect neural differences by sex often are interpreted through theories of innate sex difference. As I discuss in the next chapter, feminists have shown the fundamental contradiction between the brain's plasticity and its neural sex to be rooted not only in the intransigence of biological determinism but also in the heteronormativity of scientific models and practices.[10] Feminist responses range from skepticism of biological explanations of sex/gender, to methodological critiques, to neomaterialist rereadings of neuroscientific data. In this book I am concerned with what sort of critical perspective can respond not only to the politics of biological determinism but also to the corporeal politics of plasticity.

Critiques of science developed in the late twentieth century cast doubt on empirical claims about the biological body and refuse the taken-for-granted status of scientific facts, describing them instead as normative frames of seeing the world. Drawing from structuralism, social constructionism, and poststructuralism, often in dialogue with feminist thought, they describe the ability of social institutions, discourses, practices, and norms (over and above biological processes) to shape experience, to inscribe the body, and to produce subjects. In some respects these insights are more salient than ever. The acknowledgment that human experience is culturally and historically diverse is vital to counter the universalizing claims that are still made by the neurosciences. The recognition that difference and inequality are socially organized—"made, not born"—is needed to contest assertions that they are rooted in evolution or fixed in brain structures. And attention to how scientific knowledge is historicized is necessary to challenge both the naïve empiricism in public culture, as well as the less naïve, but nonetheless uncritical, empiricism in much research and scholarship.

In other respects, however, these critiques are no longer fully adequate to the task. Asserting the importance of nurture over nature, culture over biology, or representation over materiality has lost its critical purchase. Seeing biology as irrelevant for understanding the social, for example, underestimates the effects of experience on biological bodies, as well as bodies' own material agency. A view of subjects as cultural *rather than* biological or even neuronal, as some have argued, preserves the separate spheres of mind and brain, the physical body and the experiential one. The idea of culture as primarily symbolic and transmitted through communication (Geertz 1973) can obscure the role of bodies in generating and maintaining social structures and meanings. The claim that biology is constructed through scientific representation can conceal the ways matter acts and even participates in its own measure (Barad 2007). Ultimately, divisions of discourse and materiality, culture and nature can misjudge the stakes, which are not solely representational. As I put it elsewhere: "brain knowledge is not simply shaping what we think brains are, but is informing practices that literally, materially shape them" (Pitts-Taylor 2012a).[11]

There's more. The biological sciences are changing considerably, rendering many of the critiques of the late twentieth century less effective. Haraway (1997) writes that science tells many stories, biological determinism being only one of them. Her point is more and more evident. At the risk of overstating the case: In the postgenomic era, the sciences have enacted paradigm shifts that treat the social world as formative for the biological, see organisms as reflecting the diversity of their environments, and understand history and biology as coconstituted.[12] The idea of organisms as fixed, static, and determined by natural selection is being challenged in many realms. Rather than unfolding from genetic blueprints, organisms are being understood as participating in their own self-organization. Rather than singular and universal, they are being described as physiologically and morphologically varied through epigenetic and plastic processes. Rather than working in isolation, organisms are being situated in their environments and understood as part of dynamic systems. In some instances there is only a tinkering with biologically determinist and reductionist ideas. In others, however, there is a "more revolutionary" rethinking of the coconstitution of the biological and social. There is, then, the potential in the biological sciences for a "more complex, intersectional understanding of life" (Weasel in press, 2016). The conceptualization of the brain as social, plastic, and embodied potentially holds this promise. When brain matter can be understood as complex and multiple, rather than determined and determining, the prospects for positive theorizing are enlarged. This does not mean there is no longer need for critique—in fact, I argue precisely the opposite.

Historically, biological determinism and reductionism rendered some bodies and persons more susceptible to pathologization, erasure, or governance than others. As they are overcome by novel paradigms that propose the plasticity, dynamism, and sociality of biology, do such differential vulnerabilities persist? In the field of epigenetics, which takes account of the effects of the environment on the action of genes, Sarah Richardson shows that "certain bodies or spaces"—in particular, female, maternal bodies—are amplified as vectors of risk and "become intensive targets of intervention" (2015, 210). That is, the burden of epigenesis is distributed unevenly. In this book I focus on empirical and theoretical research programs in the neurosciences that, like epigenetics, presuppose the interplay between nature and culture, the coconstitution of the social and biological, and the situatedness of biological processes in the body and its environs. They contest mechanistic, genetically determined, and asocial ideas of the brain and the neurobiological body. They allow nurture into nature. In principle they have the potential to speak to the complexity, variation, and multiplicity of experience, rather than merely affirm social norms or justify the status quo. But the question is, how complex and multiple do they allow the neurobiological body to be? What vulnerabilities persist, and what new ones emerge?

Corporeal Politics

To understand bodily (and nervous) morphology, development, and function, scientific approaches have traditionally treated human bodies phylogenetically, as representative of a species, explaining differences between them in essentialist and reductionist terms, or seeing variances as pathologies.[13] Many social theories of the body have critiqued this universalizing, normalizing lens. They see the human body in the first instance as highly diverse, with regard to how it is lived, experienced, represented, managed, and produced. For example, feminists address how bodies are imbricated in intersectional relations of power, including race, sex/gender, sexuality, and class; sociologists of the body likewise examine how cultures make sense of, inscribe, manage, or inhabit bodies; disability scholars challenge the ablest assumptions about the body in culture, philosophy, and medicine. On the whole, these literatures understand the "body" as multiple; there is not *a* body as such, but rather bod*ies*, which are historically situated, socially stratified, and differentially experienced.

Accounting for this multiplicity has, until recently, depended on a tendency to divide the material body from the social one. Whereas the material body has been treated as the purview of the biological sciences, to be studied empirically, the social body often has been explored through the analysis of texts and practices that define, manage, discipline, or represent bodies and imbue them with value. Whereas naturalistic paradigms have historically treated the biological body as providing a presocial basis for "the superstructures of the self and society" (Shilling 2003, 37), and as responsible for determining the capabilities and constraints of human subjects, in social paradigms the body has been seen not as structuring, but structured. For example, scholars influenced by poststructuralism argue that discourses animate the body, allowing it to become meaningful in and through its discipline or regulation, or to materialize through its repeated iteration of social norms.[14] This focus on representation, even when it involves the physical body, can have the effect of eliding the material and biological dimensions of embodiment.

The case for greater physical realism is now being widely articulated. For example, feminists writing about illness, pain, and cognitive decline argue for dealing more directly with physical bodies and processes.[15] Feminists and queer writings on phenomenology and perception, public feelings, and affect describe how physical bodies are not just marked by, but also materially participate in, gendered, sexualized, and racialized relations of power.[16] Sociologists following Bourdieu (1984) insist that class and other differences are not reproduced symbolically but through physical practices and embodied cognition (Lizardo 2007; Wacquant 2015). Some disability and transgender scholars argue that purely social or discursive models of difference do not adequately grasp the fleshly and lived realities of bodily variance (e.g., Siebers 2008; Baril forthcoming).[17] To capture the experience of having a material-semiotic body, a body that is both materially real and symbolically shaped, Tobin Siebers uses the term *complex embodiment*. The theory of complex embodiment sees "the body and its representations as mutually transformative. Social representations obviously affect the experience of the body . . . but the body possesses the ability to determine its social representation as well" (2008, 25–26). My project here might in part be described as an argument for complex and *complexly embrained* embodiment.

In feminism, the treatment of the biological and social as separate spheres has been both commonplace and controversial. Recently, though, the coconstitution of nature and culture has become a key focus of feminist theorizing.[18] Feminist materialism (or "neo"-materialism) explores biology as both agentic and entangled with social meanings and cultural practices.[19] For example, feminists engage with quantum physics, dynamic systems, biology, epigenetics, neuroscience, evolutionary theory, and other scientific fields to challenge modernist, mechanistic views of matter.[20] The modernist vision sees biological matter as fixed, predictable, governed by natural laws that are incontrovertible, and producing effects that are the inevitable outcomes of such laws. By contrast, neo-materialism embraces a conception of matter as deeply situated and dynamic. It sees biology as "an open materiality, a set of (possibly infinite) tendencies and potentialities that may be developed, yet whose developments will necessarily hinder or induce other developments and other trajectories" (Grosz 1994, 191). New materialist readings of neuroscience describe not merely the influence of culture on neurobiology but also the immanent multiplicity of neural matter itself, its refusal to be fully predictable (E. Wilson 1998, 2004, 2010, 2015).[21]

The neurobiological body is now being put to many tasks in contemporary social theory. Scholars turn to the brain and nervous systems to explain affect and intercorporeality, to contest intellectualist and representationalist accounts of the self and the subject, and to elaborate biological matter as both biosocial and agentic. They draw from certain research programs in neurocognitive science and naturalized philosophy, especially those that foreground the brain's relationality with the rest of the body and the environment, extend the boundaries of the individual brain/body, or highlight its unpredictable or probabilistic capacities. Affect theorists cite neuro-scientific accounts of preconscious cognition and emotion to address the ability of the body to transmit information or feeling without need of repre-sentation.[22] They use neuroscientific work on emotions, which treats them not as propositional attitudes but rather bodily states, to ground an idea of affect as prepersonal, occurring before or below consciousness. Some also cite the neuroscientific literature on mirror neurons to suggest that social attunements or awareness of others can work at an automatic bodily register, rather than through intellectualist deliberation.[23] Neurocognitive processes may allow bodies to act outside of or before the medium of ratio-nality and subjectivity. Some argue these mechanisms also allow power to work through the modification of bodies rather than the transformation of ideas and identities.[24] Preconscious, automatic cognitive capacities are also the focus of sociologists of the body, such as those who seek to empirically explain *habitus* as the effect of bodily practices and habits rather than dis-cursive norms.[25] Here the neurobiological body is understood as a site for the generation of intersubjective awareness, the medium of perception and understanding of others, and a conduit of intercorporeality.

These efforts suggest that the neurobiological body can be a rich site of social and biosocial theorizing, but they are very divisive. Critics of the neuroscientific turn accuse scholars of appealing to the brain sciences to authorize their work while ignoring incompatibilities (e.g., Blackman 2012; Hemmings 2005; Leys 2011). Proponents dismiss these reproaches as anti-materialist, ignorant about biology, or simply wrong about the current state of the sciences.[26] I do not review these debates at length here, but my own position is that a material-semiotic view is essential to more fully grasp bodily reality. However, materialism does not absolve the need to critically access neuroscientific knowledge and practice. Neurobiology is both a le-gitimate ontological problem and also an epistemological one; what's more, these problems are interrelated.

It is not only possible but also necessary to question epistemic claims and at the same time invest in materiality.[27] Since they are not value-free or extricable from their objects of study, the neurosciences cannot be taken as a neutral lens through which to see the brain and nervous systems. It is not merely that there are epistemological biases to be eradicated; it is also that scientific practices enact their objects of investigation and have material effects on the bodies they study. I argue that we need to look more closely at knowledge of plasticity, sociality, and embodiment to see what *kinds* of brains and bodies are being enacted, erased, and transformed in neuro-scientific thought and practice. But the aim cannot be merely to debunk facts or perform negative critique that would disallow any kind of realism at all (Latour 2004).[28] The stakes, by many of the accounts discussed here, are not only symbolic but involve the way "problems are framed, bodies are shaped, and lives are pushed and pulled into one shape or another" (Mol 2002, viii). Thus in addition to thinking about representation, we also need to try to say something about ontological realities, about corporealities.

In chapter 1 I take up Karen Barad's theory of agential realism as one mode of grasping ontology and epistemology together. My broader strategy is to highlight the multiplicity of neuroscientific phenomena, including neuroplasticity, mirror neurons, the oxytocin system, and other objects of neuroscientific study. Keeping in sight their multiplicity—the different ways they are known and enacted in neuroscientific practice—can help to correct their generic treatment in social thought. It can also help to clarify their social and political implications. For example, in principle, neural plasticity allows us to describe the brain as *becoming*, always in process, never fixed or final.[29] But plasticity is not best understood as a generic principle. Rather, its specific enactment in neuroscientific research yields multiple definitions and dimensions of plasticity. I show that neuroscientific research distributes plasticity unevenly across brain regions, developmental timescales, and neurocognitive systems. One effect of this differential distribution, I argue, is that it allows the calculation of certain bodies-at-risk in ways that can both reflect and reinforce social stratifications such as class and race. This isn't only an epistemic issue; it also concerns how social institutions are governing—literally shaping—neurobiology. In research on experience-dependent plasticity and in other neuroscientific research programs, I explore how corporeal politics appear, both as forces that affect brains and neurobiological bodies, and as frames that make them visible in particular ways.

The Brain's Body

Sex difference research, as I discuss in chapter 1, reveals that there are fundamental difficulties with conceptualizing and measuring human sameness and difference in the brain. The recognition of neural plasticity does not ameliorate these problems. I explore this by looking not only at sex/gender research but also at research on the effects of poverty on the brain. This work assumes the brain is not only plastic but also situated in social inequalities. However, it risks essentializing social categories and fixing them in the developing brain—especially in the brains of those who are most vulnerable to scrutiny, such as those of minority and poor children. It also obscures how neuroscientific practices have an effect on (are part of) the phenomena they seek to measure (Barad 2007). To conceptualize social structures, neurobiological bodies, and neuroscientific measurement together, I address the plastic brain as materially performative.[30]

In chapter 2 I examine the resonance between feminism and naturalized (and neuro-) philosophy in theories of the embodied mind. Both literatures challenge dominant accounts of neurocognition as disembodied and abstract, and draw from pragmatist and phenomenological ideas of engaged, practical experience as the basis for perception and knowledge. By insisting on the epistemic significance of embodiment, they each treat cognition (or the mind) as situated. But in naturalized philosophy, theories of embodied cognition often try to account for epistemic universals. This focus ignores the ways in which social differences can generate cognitive and affective dissonance—as writings in feminist, queer, disability, and postcolonial scholarship attest (Hemmings 2012). These critical literatures have openly wrestled with the problem of essentialism while insisting on the inadequacy of universalized epistemic claims. I argue for recognition of embodied multiplicity, including cognitive and affective "misfittings" of body-minds and worlds. I borrow from Rosemarie Garland-Thomson (2011) the idea of misfitting, which describes the problems that can occur when variant bodies meet constraints in the built world.[31]

Misfittings also occur between people. In chapter 3 I address mirror neuron research, which seeks to explain the human capacity for intersubjectivity and empathy in terms of automatic, neurophysiological processes that occur before higher-level cognition or propositional thinking. In the dominant hypothesis, called embodied simulation theory, mirror neurons enable theory of mind, or awareness of other people's intentions, by registering others' motor actions with the same neural mechanisms we use for our own actions. Similarly, through simulating another's emotive expressions, mirror neurons are thought to enable a preconscious, bodily experience of others' feelings, or empathy. Mirror neurons can be understood in other ways—as learned or skilled elements of enactive perception, for example. By comparing multiple accounts of mirror neurons, I show how the dominant thesis leans heavily on a set of claims about human embodiment: that human bodies have shared motor schema, similar relations to objects, and shared phenomenological experiences of the world, all of which facilitate the automatic transfer of intersubjective information about what the other is intending or feeling. This model is influential in affect studies, but mirror neuron research is enormously controversial because of its overreach (Hickok 2014; Kilner and Lemon 2013). Rather than embracing or dismissing mirror neuron research out of hand, I carefully consider its suppositions about the body and embodiment. I use a grave example of racism and police violence both to underscore the need to address theory of mind failures and also to insist that embodiment is not inherently unifying. As one way to contest the generic, universalized social brain, I discuss research on how learned differences are thought to shape neural responses in mirror neuron systems.

In chapter 4 I explore biosocial theories of attachment and kinship. Social neuroscientists address kinship through biologically rooted, affective feelings of attachment. They argue that humans and other mammals are able to experience sustained social bonds through the involvement of neural processes linked to affect and memory. They draw heavily from animal studies of the neurohormone oxytocin, including groundbreaking research on voles conducted in the 1990s. Like other biological stories of kinship, the social neuroscientific account is closely tied to reproduction and, for the most part, focuses on heterosexual partners and mother–infant relations. Some versions are relentlessly and unacceptably heteronormative. If some social neuroscientists are getting kinship wrong, as I argue they are, this is not because they look to the biological body, but because they recognize only some bonds as biologically real and disallow or ignore the materiality of bonds that do not follow heteronormative patterns. Contesting the misrecognition and erasure of nonheteronormative bonds, I consider how kinships can be affective, biologically imbricated, and also queer. Most important, I argue that taking neurobiology seriously while insisting on its multiplicity can transform what it means to be biologically related. Thus, even while I engage in feminist and queer critique of social neuroscience, I also attest to the value of its material perspective for understanding kinship, attachment, and belonging. It is here I most strenuously argue for positive theorizing of the neurobiological body in order to more fully recognize the stakes of its social regulation.

Neuroscientific practices and styles of thought are not monolithic. Embodiment, for example, can be understood as a universal human condition that provides roughly the same constraints and capacities for all, or it can be pursued in terms of experiential, phenomenological, physical, and technological variation. Neuron systems can operate as fixed, automatic, and precognitive, or they can be symbolically modulated, and therefore affected by social differences. Neurohormonal systems can work in the service of monogamy and reproduction, or they can be understood as potentially queer. Reading these research programs diffractively (Haraway 1997) with critical literatures on bodies and embodiments, such as those found in feminist, queer, and disability studies, can show "patterns of difference that make a difference" (Barad in Dolphijn and van der Tuin 2012, 49). I hope not only to make explicit the corporeal politics at stake but also point to the "different realities and unforeseen possibilities" that can emerge (Timeto 2011, 167).

I make my way from the brain as a plastic and biosocial organ to its dependence on the rest of the body and its embeddedness in the world. I also move from the individual to the intersubjective brain/mind, and from intersubjectivity to intercorporeality. I address a number of different empirical and theoretical research programs in the neurosciences and naturalized philosophy, some only passingly and others in lengthy detail.[32] I argue for ways of looking at the neurobiological body that neither presuppose universality nor overlook the pitfalls of addressing difference. I argue, in other words, for the multiplicity of the neurobiological body and the specificity of embodied lives.

CHAPTER I

The Phenomenon of Brain Plasticity

When philosopher Catherine Malabou (2008) asks, *what should we do with our brain?*, she joins a chorus trying to awaken us to the discovery of the brain's lifelong plasticity, its ability to change and be changed. Neuroscientific ideas of brain plasticity have existed for over a century, but they were mostly confined to the development of very young brains, to learning and memory, and to recovery from injury. Otherwise, the human brain was, for much of the twentieth century, understood to be biologically determined. It seemed to be faithful to genetic blueprints, governed according to immutable rules, and, after very early development, fixed for life. One can still find many references to such a brain: in the neuroscientific literature on sex difference, for example, which leans heavily on evolutionary logic to explain the existence of brains that are purportedly organized as male or female. Yet according to many interlocutors of brain science, the twenty-first-century view is that the brain modifies itself in response to experience throughout the lifespan. The plastic brain is ontogenetically shaped in dynamic relation with its environment; this means, in the language of Bergson and Deleuze, that brains are biological *becomings*, always in process, always open to transforming themselves and being transformed. Lifelong neural plasticity may open up possibilities for agency and freedom. It may afford what Andy Clark calls a "profoundly embodied" agency (2007, 265), one that allows us to transform "who and what we are" (2004, 34). Malabou says plasticity renders us "precisely in the sense of a work: sculpture, modeling, architecture" (2008, 7). What sort of work, then, is it?

Figuring Plasticity

Brain plasticity animates naturalized philosophy as a biological condition to be reckoned with, if not celebrated. But it can also be seen as a trope of the contemporary social order, or as a justification for biotechnological intervention into everyday life. Malabou notes a resonance between neural plasticity and the demands for constant flexibility, multitasking, and self-alteration in late capitalism. Similarly, Emily Martin describes a manic plasticity demanded in the global marketplace, which asks us to be "always adapting, scanning the environment, continuously changing in creative and innovative ways, flying from one thing to another, pushing the limits of everything, doing it all with an intense level of energy focused totally on the future" (2000, 578–79). Others note that the popular discourse on brain plasticity encourages subjects to physically modify themselves with the help of neuroscientific expertise (Kraus 2012; Pitts-Taylor 2010; Schmitz 2012; Vidal and Ortega 2011). And while brain plasticity is conceived as having tremendous clinical potential for reversing the effects of traumas and degenerative diseases, it equally underpins biotechnical, pharmaceutical, and military industries aimed at cognitive modification and enhancement (Moreno 2012). A whole range of techniques, for example, involving cognitive exercises, brain–machine interfaces, drugs, supplements, electric stimulators, and brain mapping technologies, now target the brain for modification and rewiring. In the contemporary understanding of plasticity there may be no less than a "new master narrative of changing the brain-body, which thrives on the technoscientific ambition to monitor, control, and transform processes of life on the very level of their material composition" (Papadopoulos 2011, 433).

Plasticity may offer a reprieve from biological determinism, but its "dual association" with freedom and control (Papadopoulos 2011) must be confronted. How can its promise be understood when plasticity so neatly coincides with dominant ideologies and practices, or when it threatens the body-subject with techniques of governmentality? Malabou's answer is to distinguish the flexibility demanded by technoscientific capitalism from the "true" plasticity given by nature. Flexibility, she says, is a discourse that produces the subject in accordance with her neurobiology in only a distorted and superficial way. Plasticity, by contrast, is an ontological condition generated by the capacities of biology. Whereas flexibility presents an endlessly polymorphous, ultimately "reproductive and normative" subject (2008, 72), true plasticity is more rebellious. It refers to the brain's literal fashioning and refashioning, and contains a tension within itself, culminating in a "disobedience to every form, a refusal to submit to a model" (2008, 6). I find the divisions Malabou makes between epistemology and ontology (and between normativity and freedom) far too neat. However, they highlight the two matters of concern in contemporary discussions of the plastic brain that I want to address.

Construct, Property, or Phenomenon?

The first concern is whether the essence of neural plasticity can be extracted from how it is represented in science and everyday life. Naturalized philosophy often treats empirical research as more or less neutral information that can ground theories about human essence or experience. Malabou, for example, forswears any truck with neuroscientific knowledge by simply declaring herself a materialist, a stance that disallows even "the least separation" (2012, 212) between the brain and mind. Her assumption is that if one is a materialist, one must accept the neurobiological facts. If there is something wrong with neuroscientific knowledge, it is because the facts are being distorted or appropriated. By contrast, social constructionists see such representations as inextricable from the facts themselves. Even though they don't usually deny the existence of material realities, social constructionist arguments treat scientific objects as theorizable only in representational terms, as the effects of discourse. Cynthia Kraus, for example, asks not what we should do with the brain, but rather "what kind of social order and conceptions of human agency are being co-produced through knowledge claims about brain plasticity?" (2012, 253). In her view, appeals to plasticity, which are found not only in naturalized philosophy but also, as I discuss later, in feminist empiricist treatments of neuroscience, reproduce the cerebral subject who is defined by and reduced to the brain. Kraus argues that feminists should rethink whether plasticity is the "right tool for the job" (251) of combating biological determinism.

On the one side, plasticity is a biological property that is described in neuroscientific research but is essentially untouched by its representations. This stance allows Malabou to theorize the biological body despite its epistemic mediation, but it is philosophically unsatisfying. This is because it contradicts the most striking insight highlighted in her account of neural plasticity: to *think something changes the thing that thinks it*. The strict divisions between knowledge and essence, representation and being, break down in the contemporary plastic brain. On the flip side, treating plasticity as a social construct allows one to grasp how representations of plasticity are productive in and of themselves, and recognizes power/knowledge at work in descriptions of the brain. But it stops short of acknowledging how meanings are materialized *in matter*, how they literally modify brains and body-subjects, and, conversely, how they are touched by what they represent. To dismiss plasticity as a mere trope, one that has only a representational reality, foreshortens a grasp of its deeply biopolitical character. Biopolitics "has crossed the epistemic threshold" (Vatter 2009, n.p.); it involves not just the description but also the governance of biological life (Foucault 2009).

One does not necessarily have to choose between these positions. Diana Coole and Samantha Frost claim that one can "accept social constructionist arguments while also insisting that the material realm is irreducible to culture or discourse and that cultural artifacts are not arbitrary vis-à-vis nature" (2010, 27). With respect to plasticity, this means acknowledging that even though there is only mediated access to brain properties, the properties themselves must be addressed nonetheless. If discourses change body-subjects, they can only do so because bodies are amenable to being changed. Both epistemology and ontology matter; what's more, it is their *relation* that matters most with respect to the plastic brain. I argue that plasticity demands such an onto-epistemological approach, one that takes questions of being and knowing as inseparable. Karen Barad's theory of agential realism (2007), for example, argues for seeing scientific objects and their measurements together as comprising phenomena, which are both real and actively shaped. In this chapter I think of plasticity as such a phenomenon. It is a set of materialities that demands interpretation; in other words, it has ontological import, but also bears the imprints of its observation.

Whose Work Is It?

The second concern I want to address is whether and how the brain's plasticity translates into agency. In what sense, precisely, is the plastic brain a work? And, more important, whose work is it? There are a number of possibilities, including (but not limited to) (a) that the agency of plasticity belongs to the subject, (b) that it inhabits the biological body, or (c) that it lies outside the body-subject entirely, for example, given to culture. The first possibility, embraced by many popular guides to brain science, is that the potential of neural plasticity is available to a subject who can modify her own brain. Jeffrey Schwartz and Sharon Begley, for example, champion the possibility of "self-directed neuroplasticity" (2002, 254). One can alter the availability of neurotransmitters through antidepressants and other drugs; in the current age one is also told that it is possible to facilitate more efficient synaptic connections, create new pathways, and even promote the growth of new neurons. Modifications of the brain may address health problems, such as mental illness, infertility, obesity, or dementia, to name a few examples, or enhance cognitive performance. Rose and Abi-Rached (2013) describe such practices as a contemporary form of somaticization, where the individual is pressed to optimize and enhance her biology as a matter of personal management, wellness, and neoliberal citizenship. While there are many ways in which individuals can now participate in transforming their brains, the claim of self-directed neuroplasticity raises a number of problems. It presupposes a version of dualism, not quite between mind and body, but between the brain-as-mind that directs the action and brain-as-body that is acted upon. It presumes personal ownership over the brain's capacities, as if the brain responds to (and only to) the desires of its body-subject. It sees agency as a "fixed human property" (Malafouris 2008, 23) under the command of an exclusively human subject.

Those who (like me) put far less stock in the intentional subject as the master of her own biology nonetheless may find potential in plasticity. In many accounts of plasticity the brain makes itself on its own, without guidance or permission from a knowing subject. It changes without the subject's consciousness of, or control over, when or how it is being changed, and according to certain models of cognition it does so without need of symbolic representations or meanings. The brain that changes itself may be an instance of the "generativity and resilience of material forms with which social actors interact, forms which circumscribe, encourage, and contest their discourses" (Coole and Frost 2010, 26). Rather than freedom *from* biology, the plastic brain may suggest that biology itself entails a kind of freedom, which is found in the multiplicity of its potential and the unpredictability of its actualization. Deleuzian readings of synaptic plasticity, for example, lend the brain a radically agential character, even while rendering it vulnerable to control society. It is variously described as a "reservoir of potential" (Hauptmann 2010, 20); one that has a capacity to "explode its form" (Malabou 2008, 12); as having an "outsider" status, as "anomalous" and "unintelligible" (Watson 1998, 42); as engaged in conditioned reflexes but also "creative tracings" that are entirely new (Murphie 2010, 8); as entangled with technologies that allow body-subjects to "re-configure our neural connections all the time," rendering us multiple, able to do and be many things at once (Rotman 2000, 74).

Rather than ontogenetically unique, creative, and unpredictable, the plastic brain can be also seen as habituated, imprinted by the social patterning and regularity of experience. The plastic brain, according to some of the research discussed below, is not indifferent to its surroundings but rather inextricably dependent on them. Despite the temptation to read it primarily in terms of biological freedom, one can also see in neuroplasticity the brain's vulnerability to its environment, its exposure to and situatedness in the world, including to language, values, and social structures. For some observers, plasticity means not freedom and agency but rather cultural inscription, that "from birth on, our mind as well as the correlated brain structures are essentially shaped by social and cultural influences" (Fuchs 2005, 115). This may mean that the plastic brain is susceptible to social hierarchies and inequalities, perhaps even expressed in phenotypes such as a gendered or classed brain. In feminist writings the plastic brain resists biological determinism, largely through its openness to cultural shaping and influence, including gender socialization (Schmitz and Hoeppner 2014). For feminist empiricists, evidence of brain plasticity is a resource to critique scientific sex/gender bias and an alternative explanation for findings of sex differences in the brain. Although feminists are highly critical of neuroscientific claims of bifurcated sex difference, some suggest that the brain's vulnerability to gender training may explain observable differences in brain function and structure. They argue that social forces, rather than evolutionary ones, are the cause of neurobiological difference.

The move to read plasticity through the lens of gender socialization, alongside efforts to discover a neural phenotype of poverty and other biosocial research programs, underscores the necessity to theorize plasticity not as a social construction, nor as an unmediated matter of fact, but rather as a jointly ontological and epistemological concern, one whose agentic implications are not immediately straightforward. I make this case below, but first I offer a brief (and necessarily partial) description of neuroscientific research on plasticity, which suggests that it cannot be understood as generic or monolithic. Plasticity is neither an undisputable fact with a singular meaning nor a mere social construction. Rather, it refers to multiple materialities that are entangled with specific research questions, practices of scientific measurement, and ideas about development, environmental context, and biosociality. The promise of neural plasticity depends in part on how it is defined and measured.

Plasticities of the Brain

Plasticity has some shared meanings across different knowledge sites, but also gains significance and weight in particular contexts (Jordan-Young 2014; Kraus 2012; Pitts-Taylor 2010; B. Rubin 2009). At the start of modern neuroscience, the concept of plasticity emerged to address how neurons' connections with each other are related to the brain's activity. In the mid-twentieth century, this synaptic or "functional" plasticity often was elaborated in contrast to the apparently fixed structural organization of the brain. Evidence of the mature brain's ability to rewire and reshape itself in response to new stimuli and activity has more recently led to biosocial models of brain structure as well as function. This history should not be conceived in a teleological fashion, where the brain is merely awarded greater plasticity over time. Even in the current moment, when neural plasticity is more broadly recognized than ever before, the brain does not appear to be globally or monolithically plastic. Rather, in different research programs plasticity is unevenly distributed across developmental time scales, various regions of the brain, and even potentially between persons.

Habit, Learning, and Synaptic Plasticity

The term *plasticity* was used in eighteenth-century materials science to describe the malleability of matter, and in the nineteenth century to denote the ability of organisms to change in response to environmental demands (Berlucchi and Buchtel 2009). Early scientific conceptions of neural plasticity seem to have some relation to both meanings of the term. William James (1890), for example, noted that all matter, including nervous tissue, changes structure in the face of a "modifying cause." He defined *plasticity* as the possession of a structure weak enough to yield to an influence, but strong enough not to yield at once. Matter changes, and it resists change; James argued that the dual ability of neurobiological matter to both modify and stabilize, in relation to the behaviors of persons, explains why people develop habits or characteristic propensities.[1] James assumed a hydraulic model of the nervous system, which allowed him to consider plasticity in broad physical terms. With the establishment of the neuron theory, later known as the neuron doctrine, the brain's plasticity was tied to the synapse.

The neurons that Ramon y Cajal, an anatomist at the University of Barcelona, described in 1887 were unlike other cells in the body, having two different kinds of "processes" (dendrites and axons) extending from them, with points of near meeting between the axon of one cell and the dendrite of another.[2] These synapses opened up the question of how connections are made across them, and to what extent such connections are stable or can change.[3] Cajal's student Tanzi hypothesized that the strengthening of synaptic connections is linked to the consolidation of memories and the learning of motor skills, whereas Cajal proposed that the brain, at least early in life, could strengthen neuronal connections with mental exercise (DeFelipe 2006). He thought this could explain "the great intellectual capacity of certain individuals and how an individual with a small brain could become a genius" (812–13). There was considerable interest in synaptic plasticity in Cajal's time, but the topic was then abandoned for much of the first half of the twentieth century. Enthusiasm was dampened in part by fierce debate on how the connections between neurons are made in the first place, which was unresolved until the 1950s.[4]

A number of conceptual proposals midcentury renewed interest in synaptic plasticity as the mechanism of learning and memory. These include Donald Hebb's 1949 theory of associative learning, which argues that when two stimuli occur together, the near-simultaneous firing of cells results in a strengthening of their connectivity, so that they are more likely to fire together again. Conversely, the connection weakens when they fire separately. Terje Lomo at the University of Oslo later hypothesized that connections could be strengthened for long periods of time, a process he named long-term potentiation. Lomo and Tim Bliss (Bliss and Lomo 1973) tested the thesis on the hippocampi of live, anesthetized rabbits in the early 1970s, finding that stimulating a pathway with an electrode repeatedly modified its strength for hours on end.[5] Synaptic plasticity suggests that the brain can change and adapt as a function of learning while remaining structurally stable. It is this sort of plasticity that animates much contemporary social theory, in part because of the influence of Gilles Deleuze and Felix Guattari (1987).[6] They were inspired by the brain's ability to self-organize at this micro-architectural level, to act in a probabilistic rather than fixed fashion, as well as to connect remote areas of itself through the creation of networks (Murphie 2010). They described synaptic plasticity in terms of the brain's material potential (and its capture). But while synaptic plasticity allows the brain to modify itself micro-architecturally with seemingly few limits, there remains the question of its gross organization—for example, the relative size, density, and shape of different brain regions and the basic wiring of neural pathways that allow the brain to receive sensory information.

The dominant view for much of the past century was that basic neural circuitry is predetermined, unfolding automatically from a genetic blueprint. The cells in the retina would be connected to cells in the visual cortex at the back of the brain, for example, through intrinsically generated pathways that, once laid down, are permanently fixed. David Hubel and Torsten Wiesel's research in the 1960s and 1970s offered a striking alternative: that the brain uses external stimuli for its wiring. In one of their famous experiments on the visual cortex, they sewed one eye of a litter of kittens shut for three months; afterward they measured cellular activity in the column of cells in the visual cortex that would normally have received visual information from that eye. Those cells were considerably less active, and the kittens could see almost nothing out of the eye that had previously been occluded, even though the eye itself was perfectly normal. Hubel and Wiesel concluded that sensing light generates the electrical activity necessary to complete the writing of the visual cortex. As they put it, "in early life neuronal connections are only too subject to modification by the environment" (Hubel and Wiesel 1998, 407; see also Hubel and Wiesel 1970). This means the visual cortex is embodied and relational, dependent on both its own activity and its engagement with the world for its self-organization. Yet this research also points to the limits of developmental plasticity. The kittens whose eyes had been occluded for too long were never able to see normally; their visual cortices remained wired as if they had only one eye. Because of the apparently tight window of plasticity, Hubel and Wiesel retained a view of the mature brain as a fixed structure whose functions are irreversibly localized, or tied to specific neural pathways, for life (Clifford 1999).

Uneven Developments

The idea of a "critical window" of plasticity now uneasily coexists with a view of the brain's extended development and morphological plasticity.[7] The neuroscientific understanding of human brain development has been changing dramatically since researchers began to conduct imaging studies in the 1990s.[8] The human brain was once considered mostly mature by age five or six; it was thought to achieve this through a wave of prenatal and postnatal synaptic growth, followed by several years of pruning of unused synapses, which seems to buttress the efficacy of remaining ones. Neuroimaging researchers have claimed, however, that human brains experience a second wave of synaptic sprouting just before puberty and undergo pruning for years afterward. Imaging studies of the adolescent brain have reported changes in, for example, the frontal, parietal, and temporal lobes from the teen years into the early or midtwenties (Giedd et al. 2004; Petanjek et al. 2011; Shonkoff and Phillips 2014; Sowell et al. 1999). The expansion of developmental plasticity into early adulthood suggests that experience has considerably more opportunity to shape the brain than previously thought.

Within the neuroimaged brain, however, development has a "heterogeneous temporality" (Gumy 2014, 257); that is, different regions appear to have different developmental trajectories. The implications of this depend in part on whether, and to what degree, tasks and aptitudes are localized with respect to various brain areas.[9] If one subscribes to the localization thesis that dominates neuroimaging research, this means that the capacity for some tasks may be more plastic than others. The stratification of plasticity between regions of the brain is leading to new ways of identifying neurobiological subjects and to debates over the relative influence of nature and nurture. For example, the adolescent brain has come to be defined in neuroscientific research as marked by "disparity in maturation between the limbic and prefrontal regions of the brain" (259). The limbic system, which includes the amygdala and is associated in the neuroscientific literature with emotion, is understood to develop beginning at around age 10, whereas the prefrontal region, associated with executive function, such as self-control and planning, is purportedly in development until the midtwenties. Some researchers (e.g., Casey et al. 2008; Giedd 2004; Goddings et al. 2015; Johnson and Giedd 2014; Steinberg 2004, 2007, 2008) have presented this cerebral configuration—a more active limbic system combined with an immature prefrontal cortex—as the neurobiological explanation for the purported risk-taking behavior of adolescents. Some even describe it as the neurobiological substrate of social problems affecting teenagers, such as "unintentional injuries, violence, substance abuse, unintended pregnancy, and sexually transmitted diseases" (Casey et al. 2008, 111).

One serious critique of the adolescent brain research comes from sociologist Michael Males (2009), who objects to the presumption that adolescents are inherent risk-takers. While public health and criminal justice discourses since the 1980s have identified youth with crime, dangerousness, and unreliability, Males argues that adolescents actually show less, rather than more, risky behavior on certain measures, with lower rates of suicide, drug overdose, and overall accidents (except car accidents) than adults. He also argues that socioeconomic status (SES), not age, is the dominant factor influencing the higher rates of crime, traffic accidents, and "virtually every behavior for which adolescents are accused of displaying excessive risks" (15). When controlled for socioeconomic status, Males argues, the age disparities disappear. He says that the real risk of adolescence is the much higher rate of poverty than the population as a whole. The adolescent brain, then, can easily be considered as a social construction: background assumptions about the behavior of teenagers not only shape the interpretation of empirical data but also define the research question that the studies seek to answer. But identifying the "origin of problematics" (Spanier 1995, cited in Roy 2004, 265) in cultural assumptions does not invalidate that there are distinguishable developmental phases of human brains, or that brain development is relevant for understanding the experiences of adolescents. I see the adolescent brain as a phenomenon in Barad's sense. That is, it is neither a straightforward empirical reality that is simply observed and measured by neuroscientists, nor a mere social construction with an arbitrary relation to reality. Rather, it shows how the assemblage of ideas, practices, and brain matter can produce a *particular* neural difference, which carries material as well as symbolic import.[10]

While Males points to poverty as a social factor that is erased in neurobiological accounts of adolescent behavior, a number of social neuroscientists are trying to identify its impact on developing brains. For example, Martha Farah and colleagues argue that low SES, a sociological measure of income, occupational status, and education, which has long been correlated with poorer academic performance and IQ scores, affects particular brain systems (Farah, Noble et al. 2006; Hackman and Farah 2009; Hackman et al. 2010; Noble et al. 2005). The suspected mechanisms of low socioeconomic status on neural development include prenatal exposure to drugs and poor nutrition, lack of cognitive stimulation, exposure to lead and other toxins, and chronic stress, all of which are thought to disproportionately affect poor people. Farah and colleagues use SES as a proxy for these collective factors and claim that some "brain systems," or regions of the brain that they argue are linked to specific capacities and tasks, are more susceptible to them than other brain systems.[11] They explain this disparity as an effect of the earlier maturation of some regions of the brain compared with others, which renders them less and more vulnerable to environmental influence. The result, they argue, is a distinctive, consistent neural pattern: "childhood poverty does have reasonably specific neurobiological correlates" (Farah, Shera, et al. 2006, 169). This research program, about which I say more later, defines at-risk populations not through neurobiology alone, but by linking patterns of social vulnerability to patterns of differential vulnerability in the brain.

Stratifying the Adult Brain

The critical windows of developmental plasticity and also their stratifications by brain region and function exist alongside increasing recognition that the wiring and morphology of adult brains can be modified. Since the late 1980s and 1990s, researchers have maintained that the wiring of the somatosensory cortex, the area of the brain that receives haptic input from various parts of the body, is malleable in response to changes in activity or stimuli.[12] For example, teaching an owl monkey to use a spoon, a task that in one experiment took about seven hundred tries for the monkey to master, reportedly causes measurable changes in the area of the monkey's cortex that topographically represents (or receives information from) the fingers (see Merzenich 2012).[13] Andy Clark (2007) sees this kind of brain plasticity, in which new physical and mental equipment can be incorporated into the body schema, as a "profoundly embodied agency" that never arrests (263). "Since bodily growth and change continues," he argues, "it is simply good design not to permanently lock in knowledge of any particular configuration, but instead to deploy plastic neural resources and an on-going regime of monitoring and re-calibration" (269).

Plasticity is increasingly understood as a baseline state even for adults, in part because of neuroimaging research on experience-dependent plasticity in adult brains. Even in this research, however, the brain does not seem to be monolithically or globally plastic. For example, a series of studies by Eleanor Maguire and colleagues used magnetic resonance imaging (MRI) scanning to try to measure the effect of spatial learning on the hippocampus, a seahorse-shaped structure in the medial temporal lobe that in the literature is linked to spatial memory.[14] They chose their research subjects, London taxi drivers, because of the extensive spatial memorization required to earn a taxi license—all 25,000 streets of the city, which can take up to four years to learn. In their first study (Maguire et al. 2000), the researchers measured the volume of the hippocampus in sixteen taxi drivers and compared it with that among a control group of non–taxi drivers. They reported larger volume on average in the posterior area of the hippocampus of the taxi drivers; further, each year on the job corresponded with an increase in volume, which the researchers suspect is due to the reorganization of existing cortical circuits in response to spatial learning. A second study (Maguire et al. 2006) compared taxi drivers with bus drivers, who need to learn far less spatial information since they follow set routes; the study again found greater volume in the posterior hippocampi of taxi drivers. Maguire et al. concluded, "there may be a capacity for plastic change in the structure of the hippocampus . . . that can accommodate the spatial representation of a very large and complex environment" (1091).[15] In a third study (Woollett and Maguire 2011) comparing test-takers, the researchers measured the hippocampi of those who took the test at both the beginning and end of their studies. Drivers who passed the test showed an increase in posterior hippocampus volume, whereas those who failed did not.[16]

This research program has been hailed as evidence that the adult brain can change based on the "demands its owner places on it" (Schwartz and Begley 2002, 252), but it also suggests a number of limitations to plasticity. First, when it is understood in morphological terms, plasticity is constrained by the limited neural real estate available in the brain. Adding connections in one area seems to reduce them in others; in the initial study by Maguire et al. (2000), for example, the anterior region of the hippocampi of taxi drivers appeared to shrink in volume as the posterior region increased. Second, more connections or greater volume does not equate to greater ability. The taxi drivers in the second study, surprisingly, performed less well than bus drivers on a battery of tests that measured their ability to acquire new spatial information (Maguire et al. 2006). Finally, not everyone gets to be a taxi driver. Some of the test-takers pass and others fail (and some hippocampi increase in volume and others do not). Woollett and Maguire (2011) speculate that the deciding factor could be genetic disparities between individuals. Those who were successful, they suggest, could "have had a genetic predisposition toward plasticity that the nonqualified individuals lacked" (2113). While the developmental research cited earlier suggests that plasticity is not equally distributed among areas within a single brain, here plasticity is proposed to be a biological advantage afforded unequally to persons.

There are many aspects of plasticity research I have not addressed. But even in this very brief description, it should be clear that plasticity is conceptualized, measured, and enacted in multiple ways. While functional or synaptic plasticity as tied to learning and cognition is easily described in global and unlimited terms, the morphological or structural plasticity of the brain can seem to have a stratified economy, being unevenly distributed across various stages, regions, and even groups of persons. The proposed temporal, spatial, and genetic variability of plasticity, along with the brain's proposed selective vulnerability to various social influences, suggest that its implications for human agency are far from straightforward. A brain that expresses a universal configuration of developmental plasticity at adolescence, or one that is imprinted by poverty into a recognizable phenotype, is quite different from one that can be modified by the demands of its owner. Social constructionists might believe this diversity delegitimizes any efforts to theorize plasticity as a material reality, but as I explain further, I see this variability instead as evidence of the entanglement of matter, measure, and meaning.

Plasticity and Socialization

Any discussion of brain plasticity should recognize not only its partial and stratified character but also its absence in some research programs. This is nowhere more striking than in neuroscientific research on sex difference. Rather than a brain that constantly changes throughout life, many researchers are claiming that there are male and female brains, which are fixed as such before birth. The dominant sex difference theory, which dates to animal research from the 1960s, is that during the prenatal development of reproductive organs, brains, along with genitals, are differentiated by sex via their exposure to reproductive hormones. This theory was initially used to address the development of the amygdala and hypothalamus, areas of the subcortex that are believed to work with reproductive organs, but was eventually applied to the so-called higher cortical regions and to a wide range of functions and behaviors. It is used to explain findings of cognitive and affective differences, for example, in performance on tests measuring visual-spatial tasks, mathematical reasoning, perceptual speed, language skills, and aspects of social intelligence, as well as gender identity and sexual orientation (for a review, see Bao and Swabb 2011).[17] It is also used to understand autism as a gender disorder, essentially as the product of an extremely male brain (Baron-Cohen et al. 2005; Baron-Cohen 2009, 2010; for critique, see Gilles-Buck and Richardson 2014).

Brain imaging has reanimated sex difference research because it allows (images of) living, human brains to be compared structurally (using MRI) or functionally (using functional MRI [fMRI]). To take one example, Jessica Wood, Heitmiller and colleagues (2008a) used MRI scans to examine the relative size of the straight gyrus, an area of the medial prefrontal cortex, in a group of men and women. Based on lesion and fMRI studies, they argue that the straight gyrus is used in tasks related to "social intelligence."[18] In their study they found proportionally more gray matter on average in this area in women—10 percent larger straight gyrus on average compared with in men. Further, the size of the straight gyrus correlated positively with scores on a social intelligence test—the greater the volume of the straight gyrus, the better the score. None of the methods they used can give any evidence of origins or causality, but Wood, Heitmiller et al. suspect that the differences they measured are caused by "early in utero hormonal exposure interacting with expression of specific genes," which "facilitates the development of elaborate social cognition and behavior systems" (540). While the researchers make room for epigenetics or gene expression, they nonetheless use evolutionary theory to explain social intelligence as an inherently female trait, which evolved for the reproductive goals of "rearing young" (540).

Feminists have called into question this use of evolutionary theory, challenged the reliability and validity of such findings, and offered extensive methodological critiques (Fausto-Sterling et al. 2000; Fine 2010; Fine et al. 2013; Jordan-Young 2010; Jordan-Young and Rumiati 2012; Kaiser in press, 2016; Rippon et al. 2014; Schmitz and Hoeppner 2014).[19] In addition, they have demanded reconciliation of the theories of innate difference with evidence of the brain's plasticity, arguing that where genuine differences are found, there is no reason to assume they are innate. Instead, if the brain and the environment are dynamically integrated, then an environment that encourages different opportunities, activities, social scripts, attitudes, and personal characteristics for males and females may account for differences in behavior, *as well as* measures of their brain morphology. As Jordan-Young and Rumiati (2012) put it, "given pervasive gender socialization and widespread gender segregation in occupation and family responsibilities, it is utterly predictable that we would observe group-level differences between men and women in various cognitive functions. It is frankly somewhat surprising to us that we do not see greater differences and less overlap, and also would not be especially surprising to see more structural differences than there seem to be" (312). Thus, in the example of the straight gyrus, females could get proportionally more volume there as they grow up, as they increasingly use it in gendered tasks and styles of thought that demand more social sensitivity.[20] Not evolutionary dictates but rather the more intensive emotion work expected of girls and women would have an effect on areas of the brain that purportedly subtend this cognitive and affective behavior. By contrast, males could see less volume there because of pruning or cortical reorganization, as an effect of masculinizing socialization over time. In other words, gender roles could influence the volume of the straight gyrus, rather than the other way around.[21] The plastic brain, understood as "receptive and adaptive" (Kaiser forthcoming), would be both marked with and formed by the social patterning of gender (Schmitz 2012).

As feminists writing on the subject are well aware, the use of plasticity to explain such empirically observed differences as the effect of socialization is fraught. How well do binary categories describe individuals to begin with? For example, is social intelligence really a feminine trait? Do all or even most females really exhibit more social intelligence than most males, and in the same way, and in every context? The claim that women have distinctive gendered subject formations, even if rooted in socialization rather than biology, is essentializing; it reduces the diverse and complex attributes of body-subjects to binary categories. Similar problems characterize research on the neural phenotype of poverty. For example, the categorization of schoolchildren into low and middle SES, and the generalized claims made about these samples as representative of poor and middle-class people, obscure all of the racial, ethnic, gender, sexual, regional, national, linguistic, and even economic diversity within class categories. The methodology also obscures the variations and overlaps in exposure to the very mechanisms being assumed to make the difference, such as lead toxicity, nutritional intake, and quality of cognitive stimulation. A neural phenotype of poverty, like a gender-socialized brain, assumes a homogenous and distinctive patterning of experience across a diverse group of subjects.

Recognizing this, Rippon et al. (2014) argue for alternative models that do not treat sex/gender as dimorphic, "fixed, invariant, and highly informative" (1). Instead, they argue that both sex and gender characteristics should be treated as potentially overlapping in males and females, as multifactorial, as contingent on context (including the research context), and as entangled with social structures. This means,

> Any one sample will consist of individuals with an intricate mosaic of gendered attributes, the distributions for many of which will be largely overlapping and may not differ at the group level in that particular sample. Similarly, the individuals in the sample will not have "female" or "male" brains as such, but a mosaic of "feminine" and "masculine"

characteristics. Whatever female/male behavioral and therefore brain differences are observed in that particular sample are contingent on both chronic and short-term factors such as social group (such as social class, ethnicity), place, historical period, and social context and therefore cannot be assumed a priori to be generalizable to other populations or even situations (such as the same task performed in a different social context). Each individual's behavioral and neural phenotype *at the moment of experimentation* is the dynamic product of a complex developmental process involving reciprocally influential interactions between genes, brain, social experience, and cultural context. (Rippon et al. 2014, 4–5, emphasis mine)

Rippon et al. identify many limits of applying prefigured categories of the subject to brain properties. Such categories do not monolithically describe individual traits, and they are not always applicable across persons; they are situated in contexts and entwined with other factors that render generalizations inaccurate, and, most provocatively, they are contextual, emerging differently in different situations, including experimental contexts. They argue for recognizing not only existing diversities within sex/gender categories but also their openness to future transformations. Isabelle Dussauge and Anelis Kaiser (2012a) argue that gender and sexuality must be understood as performative, or "constituted in their repeated and contextual making" (142). Further, "stable categories of gender and sexual preference have to be left behind; instead, we need to open them up and focus on diversity not only within these categories but also within individual subjects" (142).

Material Performativity

Performativity theory understands expressions of sexuality or gender (or other categories of identity) not as essential traits but rather as embodied events that depend on their continual reiteration to appear stable over time (Butler 1990, 1993, 2004). As it is commonly understood, performativity produces gender through the subject's repetition of normative discourses and practices that stylize the body.[22] Many critics have pointed out that this theory risks treating the physical, biological body as a passive, generic site of social inscription (Barad 2007; Cheah 1996; Colebrook 2000; Grosz 1994;

Kirby 2008, 2011; E. Wilson 2010). But understood in material and not just discursive terms, performativity addresses matter's constitution through intra-action in the world, its inherent relationality, and its dual ability to be transformed by and to affect events (Barad 2007). A materialist-discursive version of performativity can be used to conceptualize the plastic, biosocial brain as situated, contextual, and contingent.

In her materialist formulation of performativity, Barad (2007) argues that, correctly interpreted, performativity is not best understood as the discursive inscription of the body. Rather, performativity "shifts the focus from questions of correspondence between descriptions and reality (e.g., do they mirror nature or culture?) to matters of practices/doings/actions" (133). In Barad's reading performativity does not represent or refer to, but rather *is*, what it names. There isn't a separate cultural sphere to which the disciplined, socialized, or gendered body refers; rather, the body/world is enacted in and through the practices of embodied being and "world-ing." Bodies "are constituted along with the world, or rather, as 'part' of the world" (160). Similarly, technoscientific practices do not merely represent objects nor "reveal what is already there" (361). Instead, technoscientific practices are material-discursive meetings of various entities (knowledge systems, tools, researchers, research subjects, bodies, institutions) that call forth specific material becomings. To put it another way, they are *assemblages* of matter and meaning. A plastic brain, on this view, does not represent either nature or culture; rather, it is a specific configuration of matter and meaning that achieves itself in entanglement with the world.

An adequate theory of performativity must insist not only on the dynamism of matter—its unremitting capacity to make and remake itself, with or without human effort—but also on the specificity of its becoming, including within scientific practices. To grasp materiality not *in spite of*, but rather, *in relation to*, its epistemic mediation, Barad proposes to address the measurement as part and parcel of matter's specificity, its particular realizations in the world. Knowledge about matter, Barad says, involves a differential attentiveness to what matters. Thus, instead of speaking of scientific objects, she addresses *phenomena*. A phenomenon includes the entities under investigation, the scientific tools and practices that touch them, the knowledges that inform them, and the material changes that measures make. Neural plasticity can be understood this way. To make sense of the plasticity of the brain, scientists, scholars, and policymakers call forth particular configurations of bodies, brain matter, measurements, and other practices. I find this a useful way to think about the problem of the gendered brain, or, in the example I want to elaborate, the neural phenotype of poverty.

Poverty as a Neural Phenotype

In the series of studies I mentioned above, social neuroscientists have argued that because of their distinct patterns of experience, a particular neural pattern distinguishes the brains of poor from middle-class children (Farah, Shera, et al. 2006; Farah, Noble, et al. 2006; Hackman and Farah 2009; Hackman et al. 2010; Noble et al. 2005). This neural pattern is found in particular "brain systems," which are configurations of structure and function that have been identified through combinations of cognitive and neural testing methods. They argue that children with low SES are at risk of having an impaired language system, which they locate primarily in the left temporal and frontal areas, and, to a lesser degree, an impaired executive system, which they argue utilizes the lateral and ventromedial areas of the prefrontal cortex and the anterior cingulate. Other brain systems are relatively untouched; they found no difference in spatial and visual cognition, and (in contrast to previous findings in the literature) no difference in impulse control or the ability to delay gratification. This particular configuration of the brain comprises, as they see it, a neural phenotype of poverty. Mapping the effects of poverty in the brain does not *necessarily* fix it there. The durability of this pattern, they admit, is uncertain: Poverty's effects could be contextual, disappearing when the context changes, they could be habituated, or they could suggest "trait-like features of brain structure and function" over time (Hackman and Farah 2009, 70). Alternatively, poverty could be understood not as a characteristic of a person at all, but rather as a *circumstance* that has an impact on cognitive performance (Mani et al. 2013).[23]

There is a large literature on the effects of poverty on childhood development, health, academic success, and life prospects. To address this in specifically *neural* terms involves the generation and imaging of patterns of brain activity during cognitive tasks using specific technologies, the naming of particular brain systems, the creation of cognitive tests that can theoretically be mapped to those systems, and, if socioeconomic status is taken to be a more or less stable characteristic of a person, the categorization of research subjects by SES. In many of the aforementioned studies, the research subjects are also 'ethnically matched' so that African American poor children are compared to African-American middle-class children, with no other ethnicities represented at all.[24] This means the research is curiously both racialized (every poor subject is Black) and race-blind (only class is measured), even while other categorizations are created (see Hackman et al. 2012 on SES differences by both gender and neighborhood). There is also the conduct of experiments that test those subjects and the enactment of statistical correlations of test scores, images, and measures. These are not solely representational but also material practices, making cuts to enact particular differences.

In some studies, for example, behavioral differences between classes of research subjects are elicited through experiments using cognitive tests. In others, neural differences are elicited through technologies such as fMRI and event-related potentials (where electrodes placed on the scalp measure voltage changes coming from neurons, and averages of those measures are timed with specific behavioral tasks). In the first case, behavioral differences are sometimes interpreted as proxies for neural difference where no direct neural difference is examined, as when a test score is thought to indicate brain capacities in specific systems (Farah, Noble, et al. 2006; Farah, Shera, et al. 2006). In the second case, neural differences are deemed significant whether or not differences in behavioral measures are found. For example, some studies report different patterns of neural activity between research subjects with low SES and those with middle SES during the execution of certain cognitive tasks when there is no measurable difference in how well the research subjects perform the task (D'Angiulli et al. 2008; Hackman and Farah 2009; Kishiyama et al. 2008; Stevens et al. 2009). Another study reports neural difference without measuring a task at all (Tomarken et al. 2004). Despite behavioral parity or the absence of behavioral difference, in these studies neural disparities are deemed significant and interpreted to suggest differences in cognitive capacities. Using the logic of reverse inference, neural patterns are interpreted as evidence of less efficient use of neural resources (D'Angiulli et al. 2008; Stevens et al. 2009) or even, absent any symptoms, pathology or brain damage (Kishiyama et al. 2008; Tomarken et al. 2004). In these examples neural differences are, quite literally, made to matter.

The phenomenon also includes the institutions that help to support research programs, the social policies that researchers speak to, and the interventions that they seek to make. Although its etiology is widely understood to be economic and structural, the proposed interventions for poverty's neurobiological effects are increasingly technoscientific, targeting the brains of specific subjects, particularly poor children of color. For example, school districts and social services programs that serve low-income, urban populations with high percentages of African American and Latino students are experimenting with curricular reform and the use of computer games aimed at improving aspects of executive function such as inhibitory control, working memory, and cognitive flexibility (Diamond et al. 2007; Diamond and Lee 2011). Such practices represent a departure from traditional institutional strategies to combat the effects of poverty on school achievement.

> Until now, interventions have been targeted at changing SES directly by increasing family income, influencing the putative mediators of SES effects, such as parenting style, and influencing academic achievement and psychopathology through direct interventions, including educational or treatment programmes targeted at low-SES communities. The targeting of brain development has involved familiar approaches, such as improving children's access to medical care or nutritional supplementation. More recently, it has included programmes aimed at training particular neurocognitive systems directly, for example, by using computerized, game-based strategies for training executive functions or school curricula that employ specific exercises as well as overarching strategies to promote executive functions throughout the school day. (Hackman et al. 2010, 11)

The widespread adoption of these practices—neuroscientifically based training programs and curricula reform whose explicit aim is to ameliorate the disadvantages of poverty in the brain—may depend on targeted investments in early childhood education (see Diamond et al. 2007; Raver 2012). However, they are also compatible with neoliberal divestments in macroeconomic, public health, and community-based programs that address poverty itself.

The idea that poverty affects neurocognition fits well with a biopsychosocial model of health and disease. While poverty is economic and structural, the deprivations and vulnerabilities that can be said to mark any person's experience of poverty are embodied, felt, and often physical. For example, childhood poverty is positively correlated with elevated levels of cortisol (a marker of chronic stress), higher rates of mortality, and greater risk of chronic illness. Because they believe it makes an imprint on the brain, Farah, Shera, et al. (2006) argue that poverty is a bioethical issue, affecting people's health and biological capacities. They articulate a desire to offer increasingly precise targets and effective interventions that can help vulnerable populations, improving their life chances irrespective of whether poverty itself is conquered. But researchers commonly fail to recognize the role technoscientific practice itself plays in bringing forth specific configurations of neural difference. The cuts that are made are rooted in particular commitments and enact specific differences that are made to matter.

For example, while research on the neurobiological effects of poverty is commonly informed by broad evidence of poor children's inferior academic performance, especially language skills, many interventions are aimed not at linguistic but executive functions (EFs) and corresponding behaviors, particularly self-control. While EFs are also correlated with academic performance, they are in addition associated "with such problems as ADHD, teacher burnout, student dropout, drug use and crime" (Diamond et al. 2007, 1387). The strengthening of executive function through modifying the prefrontal cortex reportedly has societal benefits; it may, for example, "reduce needs for costly special education, societal costs from unregulated antisocial behavior, and the number of diagnoses of EF disorders [e.g., ADHD and conduct disorder]" (1388). Further, "Interventions addressing self-control might reduce a panoply of societal costs, save taxpayers money, and promote prosperity" (Moffitt et al. 2011, 2693). The choice to target executive functions and associated brain regions over other functions and brain areas is quite explicitly linked to the identification and prioritization of social problems that are entangled with race, gender, and class inequalities.

The attempted modification of the prefrontal cortex as a goal of early education, the privileging of this brain system in configuring pedagogy, and the framing of these projects as responses to social problems (such as reducing high rates of psychiatric medicalization and incarceration in poor communities of color, reducing taxpayer costs, or training a twenty-first-century workforce) reflect not just bioethical concerns. They are also strategies of what can be called *preemptive* neurogovernance (Cunningham 2012).

Since its earliest modern elaboration, in fact, plasticity has been envisioned and enacted through the modification and preemptive governance of individuals and groups. James (1890) argued in *Principles of Psychology* that we should work with the plasticity of the nervous system for human improvement. "The great thing, then, in all education," he wrote, "is to make our nervous system our ally instead of our enemy. It is to fund and capitalize our acquisitions, and live at ease upon the interest of the fund. For this we must make automatic and habitual, as early as possible, as many useful actions as we can, and guard against the growing into ways that are likely to be disadvantageous to us, as we should guard against the plague" (122).[25]

Conclusion

At the start of this chapter I raised two concerns with respect to brain plasticity that I see as intimately related. First, the brain that is recognized as plastic, or constantly capable of transformation, raises all sorts of possibilities for agency, human and otherwise. Among these possibilities, I juxtaposed the idea of the plastic brain as a *becoming*, which recognizes a nonhuman agency in matter that precedes or resists discourse, with the notion that the plastic brain is a work of culture, an imprint of collective human effort. Each view can be supported by neuroscientific research. Some research programs encourage a sense of plasticity as more or less intrinsic, others as extremely dynamic; some accounts allow the suggestion that it is potentially infinite, others finite and stratified. This does not mean that plasticity is a mere construct with no material reality. Rather, it suggests that plasticity is materially and discursively specific; it also refers to multiple neural ontologies that are brought forth in particular research programs, which themselves may affect the structures and functions they seek to measure. This leads to the second issue. The specificity of matter must be explored without accepting the naïve position that what neuroscientific research uncovers is given by nature alone. The brain isn't plastic because we say it is, to paraphrase Barad (2007), and the ways we know its plasticity do not determine them. But the differential differences of plasticity, the way its differences matter, are entangled with our efforts to know it. In this sense the plastic brain is materially-discursively performative.

In his essay on Henri Bergson, James (1909) argued that the concepts we use to grasp reality fail to reflect its fluid nature. "The essence of life is its continuously changing character," he wrote, "but our concepts are all discontinuous and fixed, and the only mode of making them coincide with life is arbitrarily supposing positions of arrest therein" (cited in B. Rubin 2013).[26] One could argue that this epistemic limitation applies to all efforts to describe the biosocial world, but it is especially paradoxical with respect to neural plasticity. Neuroscientific research on plasticity fosters an appreciation of the brain as a fluid *becoming* that constantly transforms itself in relation with the world, but it also imposes various moments of arrest. The uneven economies of plasticity allow researchers to identify distinct patterns and configurations in the brain, which are not necessarily understood to be permanent, but nonetheless gain duration in part through their influence on future trajectories. Efforts to depict brains as biosocial also rely on a parallel effort to fix persons in social categories, statistical aggregates, or populations that can be mapped onto the brain. The resultant neural phenotypes may resist biological determinism, but they do not automatically resolve the attendant problems of reification and essentialism. The measurement of human difference through biosocial plasticity can require cuts that belie biosocial complexity and that (mis)construe the experience of individuals and groups as homogenous and predictable.

The significance of this cannot be fully appreciated through a critique that addresses neurobiology *only* in terms of its epistemic constructs. Nor can it be grasped through the assumption that the materiality of the brain simply escapes its measurement and representation. James (1909) writes that concepts "are not part of reality, not real positions taken by it," and "you can no more dip up the substance of reality with them than you can dip up water with a net, however finely meshed" (cited in B. Rubin 2013). I disagree. Concepts can help you hook up children to machines, can quite literally evoke potentials, and therefore bring about particular configurations of neurobiological difference. In James's time and in the contemporary context, the meeting of the conceptual and the biological has material effects and demands not only bioethical but also biopolitical accountability.

CHAPTER 2

What Difference Does the Body Make?

PRELUDE

Embodiment—having a body, needing a body, and experiencing life in and through a body—is something you and I have in common. I think and write this book through my body, and you read it (or hear it) through yours. We each bring our embodiments to encounters with text, as well as to interactions with other persons, nonhuman animals, objects, and environments. Embodiment locates us in a space and place, while allowing us to extend ourselves with technological means. Embodiment gives us phenomenological access to our worlds and provides opportunities for action, while allowing us to modify those worlds and opportunities. It lets us perceive others and renders us perceivable, sometimes visible, to others. Having a body is agentic for us, and yet it also makes us "vulnerable by definition" (Butler 2010, 33). No matter who or where we are, having a body places practical demands on us, and gives us each an incurable condition of mortality. In these ways and others embodiment is a shared condition of human experience.

Yet because bodies are differently located in the world, undertake different practices, and are affected by social hierarchies—and because they physiologically vary as well—embodiment is a site of difference as much as commonality. I do not mean only that bodies are represented differently, marked on the surface. Specificities, peculiarities, and inequalities are part of lived, practical, felt embodiment. For every body-subject, embodiment is particular, specific, and local as well as transversed by social patterns. Between persons—and for persons experiencing different settings, temporalities, and situations—embodiments vary. Bodies diverge in their sensorimotor capacities, bodily boundaries, perceptual tendencies, and orientations toward the world. They differ in the space they take up (and are allowed to take up) and the safety they feel. They bring to encounters different skills and histories. They are variously modified and transformed by experience: They are enabled and debilitated, trained and inhibited. They are made and make themselves, in relational terms, larger or smaller, more and less attentive, more and less mobile, more and less emotionally attuned, engaged, visible. They are settled and unsettled, and their relation to objects, people, and other creatures changes and can be changed. They differ in their attachments to others and in the affordances the world offers them. They are more and less vulnerable, and suffer more and less. Their trajectories are long, or cut tragically short. They variously fit and "misfit" (Garland-Thomson 2011) in the world. In considering the embodied mind, the question I want to ask (one that many before me have asked in other contexts) is: Do these differences make any difference?

Materialism and the Rise of Embodiment

Patricia Churchland's (1986) declaration that *the mind is what the brain does*, which proclaims the necessity of the brain sciences for understanding the mind and consciousness, is both materialist and reductionist. A materialist view of mind holds that "mental properties for perception, for knowledge, for learning, for memory all had to come out of the complexity of the organization of matter" (Churchland in Campbell 1996).[1] Reductionism as it applies to neuroscience holds that the only matter that counts is brain matter; the mind and all that it entails can be entirely explained through neural processes. In its most extreme form this could mean, as Francis Crick claims, that "You, your joys and sorrows, your memories and ambitions, your sense of personal identity and free will, are in fact no more than the behavior of a vast assembly of nerve cells and their associated molecules" (1994, 3). This view, that the mind is exclusively neuronal, is contested within naturalized philosophy and neurocognitive science by the idea that *the mind is fundamentally embodied*.

The embodied mind paradigm insists that the mind is irreducible to the workings of any single organ or system. Rather, the mind is dependent on "the kinds of experience that come from having a body with various sensorimotor capacities" (Varela et al. 1991, 172), which "are themselves embedded in a more encompassing biological, psychological, and cultural context" (173). Embodied mind theories, which include theories of emotion and memory, extended cognition, and enactive perception, are materialist and physicalist, but they are also nonreductive. In contrast to the idea that the mind is wholly reducible to the brain, embodied mind theorists draw from phenomenology and pragmatism to insist the mind is dependent on, and even constituted by, the actions of the body as a whole and on the environment. Embodied mind theories remain somewhat marginal in neurocognitive science, but they have been hugely influential in the social sciences and humanities, where they inform contemporary materialist, biosocial, and affective social theories. To recognize embodiment as an inescapable feature of perception, thought, and consciousness offers a powerful rejection of neuroreductionism, while at the same time contesting rationalism, the idea that knowledge is based on abstract reason. If the mind is neither the abstract processing of symbols, nor exclusively the workings of neural networks, but rather is embodied, it can be understood only in terms of lived experience (Fuchs 2009). A mind that is embodied, potentially, can mean it is immanent (tied to the capacities and worlds in which it is enacted); relational (affected by its position to and interaction with other minds, bodies, and objects); affective (shaped by feeling and emotion); and situated (tied to specific places, needs, and circumstances).

This discovery of embodiment resonates to some degree with feminist insights. Such ideas have since the 1980s been central themes of feminist epistemologies, which have argued in various ways that "our direct embodied experience of the everyday world [is] our primary ground of knowledge" (Smith 1991, 22). Thus Miriam Solomon (2007) argues that feminist and neurocognitive ideas of the embodied mind are not merely compatible but rather can be counted as part of a broad transdisciplinary intellectual movement advocating *situated cognition*. But while they do share phenomenological and pragmatic orientations, Solomon obscures a serious rub: In feminist thought, bodies and embodiments are heterogeneous. Because bodies are differently located in the social world, and social hierarchies affect the experiences of body-subjects, embodiment is as much a site of difference as it is a site of commonality. Epistemically, embodiment in feminist thought amounts to some kind of difference, or, better, multiplicity, the idea that there are multiple truths and ways of knowing. Feminists have argued that an embrace of epistemic multiplicity is valuable not only as a corrective to scientific assumptions of objectivity but also to challenge universalizing claims about human experience. As Helen Longino summarizes, "with the embodiment of the subject, experience must be rethought, as it can no longer be understood as the parade of sense data whose character is the same for all perceivers" (2010, 734).

By contrast, naturalized perspectives commonly assume "a model of the body that is more or less universal, leaving bodies and bodily behaviors that differ from this model to be considered abnormal, pathological, or less developed in some way" (Pitts-Taylor 2008, xxii). The embodiment that often appears in neurocognitive, naturalized philosophy is tied to a generic, ideal model of the body, or its difference is treated as a special case that can put normal embodiment into relief. If embodiment is to be a solution to neuroreductionism, as is being widely proposed, I argue that embodied mind theories cannot assume the uniformity of the body and bodily experience, but rather must pay attention to discrepancies and dissonances in how minded bodies and worlds fit together. The task demands rethinking the universalizing tendencies of naturalized models of the embodied mind, while also resisting the essentializing tendencies of early feminist epistemologies. In this chapter I address these conflicts and explore what epistemic differences having a body can make. I use an assemblage theory of disability to try to conceptualize examples of epistemic multiplicity and cognitive dissonance in relation to the body. If the "history of disabled people in the Western world is in part the history of being on display, of being visually conspicuous while being politically and socially erased" (Garland-Thomson 2002, 56), the multiplicity of embodiment for persons deemed disabled is both readily apparent and also too readily ignored.

The Embodied Mind in Naturalized Philosophy

The embodied mind paradigm is a diverse family of theories that describe cognition as variously embodied, enactive, embedded, and extended, as well as emotional or affective. I introduce a few of these concepts below, which can collectively be referred to as "embodied" because each in some way presents cognition in terms of bodily experience, highlighting, for example, the body's sensorimotor capacities, action in and felt engagement with the world, or embeddedness in the environment. In these respects the theories can be seen as an extension of the American pragmatist tradition (Johnson and Rohrer 2007). Pragmatism articulates an alternative to Cartesian rationalism, the idea that knowledge is derived from reason rather than experience. Against this view, early pragmatists, including John Dewey and William James, defined cognition as practical, problem-focused, ongoing action in the world. From evolutionary theory, they discerned that cognition could not be understood outside of the bodily and environmental context. That is, "everything we attribute to mind—perceiving, conceptualizing, imagining, reasoning, desiring, willing, dreaming—has emerged (and continues to develop) as part of a process in which an organism seeks to survive, grow, and flourish within different kinds of situations" (21–22). The purpose of consciousness, then, is not the creation of internal representations that mirror external reality. Instead, consciousness "is about acting— it emerges through processes that make the world available to those systems that allow us to select behavioral means to our ends" (Prinz 2008, 434; see Prinz and Clark 2004). The pragmatists also held that the so-called higher-level functions of language and consciousness are not ontologically distinct from "lower"-level unconscious and immediate sensory experiences. The former, as Dewey put it, "grow out of organic activities, without being identical with that from which they emerge" (1938, 26, cited in Johnson and Rohrer 2007, 23). This means that there are not strict divisions between thinking, feeling, and perceiving.

Emotions are felt throughout the body—the gut, the hands, the heartbeat—and in the neopragmatic view they are not the effects of thoughts but perceptions of bodily states that shape the content of cognition. Antonio Damasio's (1996) well-known somatic marker hypothesis, for example, describes the physiological states that accompany stimuli, such as the raised blood pressure, sweating, and muscle tension that accompany fear, as marking experience not through conscious memory but through embodied feeling. These "somatic markers," he claims, are partially reenacted, or simulated, in future experiences of encountering the same or similar stimuli. This lends new stimuli somatic valence that is biographically specific. Jesse Prinz (2004) calls emotions embodied appraisals because they not only somatically mark experience but because, he argues, they in and of themselves amount to judgments of the world. As valenced apprehensions of the body's changing relation with the environment, they "represent situations as matters of concern" (74). Many social theorists have embraced such theories of emotion, arguing that they can be used to interrelate cognition with external structures that influence personal, felt experience, such as the negatively valenced experiences of stigma, and to address intercorporeal and prepersonal dimensions of experience that do not depend on representation or language.[2]

Embodied mind theories also draw from phenomenology to address cognition in terms of the enactment of perception. Enactivism depicts the mind as involving the whole body and its motor and perceptual systems, and understands cognition as actively experiencing or "enacting"—not contemplating—the world. This view has roots in the seminal work *The Embodied Mind*, by Francisco Varela et al. (1991), which uses dynamic systems theory along with pragmatism, phenomenology, and cognitive science to depict cognition as an event of the minded body "structurally coupling" with the world. Enactivism argues that cognition proceeds without the need of representation. Instead, cognition is a constant, multidirectional, intra-causal interaction between the embrained body and environment. This claim depends on a strong view of sensorimotor action as constitutively contributing to cognition. Moving about in the world, perceiving it, and acting in it provide the context and the shape of what we cognate. Alva Noe (2004), for example, describes perception not as the passive reception of stimuli, but rather the activity of unconsciously (but nonetheless skillfully) selecting and organizing the features of the world that are relevant to us. To understand what is relevant to us at any given time, enactivists point to cognition's embeddedness in the environment. For Noe, perceptual content is both factual (dependent on how things *are*) and relational (dependent on the vantage point of the perceiver). Arguing that perception is as haptic as it is visual, Noe describes the blind person who perceives by touch, "not all at once," but over time, with "skillful probing and movement," as the ideal example of enactive perception (1). This reference to disability as a model of embodied cognition needs attention.

If physical embodiment shapes the mind, what kinds of limits do bodies provide? George Lakoff and Mark Johnson's (1999) embodied metaphor theory explains how, as they see it, the motor and perceptual systems are largely responsible for the content of cognition.[3] They define metaphors as cognitive abstractions (or combinations) of brain activity in the sensory and motor systems. The sensory system brings stimuli such as light, pressure, smell, and sound taken in from the eyes, skin, nose, and ears to the brain. The motor system links the brain and spinal cord to muscles throughout the body that move one about in the world, exposing one to proprioceptive stimuli, while glands register various internal and external states of the body (temperature, blood pressure, and so on). All of this pre-discursive sensorimotor information is instantiated in the brain through neural representations, which are not representations in the symbolic sense but patterns of synaptic connections that are activated by specific senso-rimotor stimuli. The brain combines multiple neural representations to generate metaphors, which provide the basis for concepts. Cognition, then, does not involve working through a set of symbols, but rather through neural maps of our bodily experience in the world.

This view leads to *embodied realism*, where the content of a percep-tion is tied to the bodily mechanisms of its production. For instance, col-ors are neither objective, existing outside of perception, nor purely sub-jective; rather, they are a "creation of specific neurobiological capacities intra-acting with particles with specific wavelengths" (334). This version of embodied realism in some respects recalls Donna Haraway's doctrine of embodied objectivity, where the material capacities of perception con-dition what is known.[4] Both theories require us to abandon, as Lakoff and Johnson put it, "the correspondence theory of truth, the idea that truth lies in the relationship between words and the metaphysically and objectively real world external to any perceiver" (334). But unlike Haraway, Lakoff and Johnson emphasize biological constraint on thought. While neural maps are learned, a biological body, with universal properties selected by evolution, makes them possible. Because they are based on human phys-iology, the range of basic human concepts is limited: "our sensory-motor systems thus limit the abstract reasoning that we can perform. Anything we can think or understand is shaped by, made possible by, and limited by our bodies, brains, and our embodied interactions in the world" (5). Because they are neurally instantiated, conceptual systems cannot easily be changed. This is one reason why "we are not free to just think anything"; that is, "our conceptual systems are instantiated neurally in our brains in relatively fixed ways" (5).

While Lakoff and Johnson are concerned to identify the biological boundaries of cognition, Andy Clark (2008b) takes the opposite tack, broadening the mind to constitutively include the environment with which the mind-body interacts. Thus the brain, body, and the surrounding world are seen as a cognitive economy consisting of "neural, bodily and environmental contributions and operations" (217). In Clark's account of extended cognition, the brain relies on the outer world for information holding, sorting, and other work that helps it deal with its own limits of memory and attention. Further, the brain utilizes what Clark calls cognitive technology—everything from pencils and notebooks to computers—to such a degree that they become part of the cognitive system. These technologies "alter the computational spaces" (Clark 1998, 47). In his view, language, concepts, and symbols, as well as social structures, do this, and "external artifacts and social organizations likewise alter and transform the tasks that individual brains need to perform" (47). Clark argues that our dependence on cognitive prostheses breaks down meaningful boundaries between humans and machines, and organic and nonorganic systems. This cyborgian capacity depends on the brain's plasticity, its ability to be transformed through new activities, use of new tools, and new stimuli. For Clark, "human minds and bodies are essentially open to episodes of deep and transformative restructuring, in which new equipment (both physical and 'mental') can become quite literally incorporated into the thinking and acting systems that we identify as minds and persons" (2007, 264).

In contrast to embodied realism, Clark argues that the content of cognition is not determined by the commonalities of human or even organic bodies; conceivably, an infinite number of configurations of cognitive economies are possible.[5] This means that the minded self or the subject is an assemblage, a "hastily cobbled together coalition of biological and non-biological elements, whose membership shifts and alters over time and between contexts" (2004, 177). Like Haraway, Clark offers a cyborgian theory of the body-subject, but as I explain, they draw different epistemic conclusions.[6]

Embodied mind theories in naturalized philosophy cannot be read monolithically; in fact, they make distinct claims about the body and embodiment. I discuss some of these differences below. First, however, I address the proposed resonance of embodied mind theories with feminist thought. Solomon (2007) argues that neurocognitive embodied mind theories should be considered alongside feminist epistemologies as part of a transdisciplinary movement advocating situated cognition.[7] Situated cognition, according to Solomon, overturns classical ideas of cognition as general and universal, abstract, and symbolic. Instead, meaning emerges from the interaction of the minded body with its environment. Rather than abstracting what is common in all cognition, situated cognition is best suited to examine "the epistemic significance of particular routes to cognitive accomplishment" (413). And rather than reducing the mind to the brain, situated cognition minimally sees mind as dependent on the body as a whole, or maximally as a fluid assemblage of brain, body, and world. On these points the two traditions resonate. Yet in feminist thought the situated character of knowledge results in multiple truths and ways of knowing. Despite their roots in radical empiricism, I argue that some of the most prominent embodied mind theories in naturalized philosophy elide or outright refuse epistemic multiplicity, seeking instead to explain cognitive universals. A rush to assimilate neurocognitive and feminist epistemologies would obscure this key distinction. It would also gloss over the difficulties of how to articulate the specificity that embodiment brings.

Epistemic and Embodied Multiplicity

Although they are extremely (and necessarily) vast and diverse, feminist and queer writings on embodiment that address gender, sexuality, race, and dis/ability, on my reading, offer a collective sense that there are *differences and contradictions in embodied experiences* that lead to diverse epistemic outcomes. They indicate that, to use N. Katherine Hayles's phrasing, "Embodiment can be destroyed but it cannot be replicated" (1993, 91). This is not only to say that every embodied life is in some way unique, but also to suggest that the embeddedness of individuals in social patterns and historical contexts generate varied epistemic conditions and create multiple ways of perceiving (and enacting) the world. Feminist standpoint theorists, for example, adopt the phenomenological claim that embodiment shapes perception, but for them this is not a universalizing insight; instead it explains how knowledge is tied to conditions of experience *that are socially differentiated*, such as the gendered organization of labor and reproduction (Smith 1988, 1991, 1992; Young 1990).[8] Intersectionality theory describes not bifurcated but rather multiple social locations, complicated by race, class, sexuality, and other relations of power (Collins 1990, 2000; Crenshaw 1989).[9] The result of intersectionality is heteroglossic; the local and situated character of knowledge results in the multiple and conflicting nature of epistemic truths. Further, inequalities generate not only epistemic differences but also cognitive and affective dissonance (Hemmings 2012), involving a sense of the fractured, partial, and relational character of perception and knowing.[10]

These literatures also reveal the difficulty, if not impossibility, of reducing epistemic variation, embodied difference, and dissonance to prefigured categories of the subject. Standpoint theories challenge universal ideas of knowledge, but they risk essentializing gendered subjectivity (such as a "women's standpoint"), which would reify the differences between, while concealing the diversities within, genders. Intersectionality theories are similarly criticized for thinking "only in terms of existing sociopolitical categories, especially gender and race," and for codifying identities (Bost 2008, 340; see also McCall 2005).[11] At their best, these frameworks do not reify subject positions but acknowledge the mobility of knowers and knowledge positions (Puar 2012).[12] Haraway's (1988) concept of situated knowledges argues not only for the epistemic but also the ontological multiplicity of knowing subjects. In her account vision is shaped not only by the social locations of gendered, raced, classed individuals but also by the material capacities and technologies that compose knowing body-subjects. She sees knowers as cyborgian assemblages of organic and nonorganic matter (and meaning) that can change. The result, for her, is the material-discursive specificity, and therefore partiality, of all knowledge. She writes of "heterogeneous multiplicities," which are "simultaneously salient and incapable of being squashed into isomorphic slots or cumulative lists" (586).

As feminist and queer phenomenologists have argued, power relations are also felt in varied bodily boundaries, perceptual tendencies, and orientations toward the physical and social world. In circumstances of inequality the result is not merely experiential pluralism but dissonance—revealed, for example, in the "time and work [it takes] to inhabit a lesbian body" (Ahmed 2006, 564) or the feelings of shame generated by racialization (Alcoff 2006). The felt relationality of bodies—their ability to affect and be affected—also renders them vulnerable before or below consciousness or self-awareness

(Clough 2007, 2010; Puar 2007, 2012; Sedgwick 2003).[13] Feminist and queer affect theorists argue that racialized, classed, gendered, sexualized relations of power can be experienced as bodily changes that are "irreducible to the individual, the personal and the psychological" (Clough 2007, 3). For example, in her discussion of post-9/11 modes of securitization and biopower, Jasbir Puar (2007) shows how a Sikh man wearing a turban can be turned into a national threat—one that is specifically racialized and sexualized as a "terrorist fag"—virtually overnight. Racism, homophobia, and nationalism can create the queer terrorist body not as a subject or an identity, but instead as "an affective and affected entity that create[s] fear but also feel the fear they create" (174).[14] Although some theorists insist on affect's complete autonomy from signification, affect can be seen as a way into thinking about the "mutual imbrication" of the body-mind and world, or "as a kind of knowledge about the interface between ontological or epistemic considerations" (Hemmings 2012, 149; see also Ahmed 2006; Blackman 2012; Leys 2011). For example, drawing from the writings of Franz Fanon and Audre Lorde, Hemmings argues, "racially marked subjects can have [both] a critical and affective life that resonates differently" (2005, 564).

Feminist and queer accounts of embodiment, on the whole, see it as a differentiating condition, one that multiplies rather than universalizes cognition or affect. The perspectival, spatial, and temporal locatedness, historicity, visibility, and vulnerability afforded by the body is not generic and unifying, but specific and woven through with relations of power and inequality. As I discuss shortly, disability scholars also have contributed richly to theorizing embodiment as variant and multiple. The recognition of disability, whether conceived as a social construction or as a material reality, disallows universalizing treatments of the body and embodiment. Attention to bodies deemed disabled also reveal the many ways in which the environment is more and less functional for different people and in different circumstances. They show not only that bodies have varying capacities and morphologies but also that environments and social investments affect how well bodies and worlds come together. Attention to bodily vulnerabilities can also show their queerness; that is, "bodies are not just different from others—as monolithic identities—but sometimes individual bodies, in themselves, are not recognizable as singular, stable, or categorizable identities" (Bost 2008, 361).

Like embodied mind theories in naturalized philosophy, each of these feminist perspectives addresses the inadequacy of abstract, universal Reason through attention to the body. But unlike them, they also suggest that the commonality of consciousness, phenomenological experience, and affective life cannot be presumed. In fact, they insist on epistemic or experiential differences due to variations in embodied lives, which are affected by but irreducible to social patterns and inequalities. They also raise the question of alterity, the recognition of differences that "cannot be known in advance," but which "brings us to the limits of our own self-certainty and certainty about the world" (Weil 2010, 15). If cognition, perception, and consciousness are enacted through bodies, and if embodiments are heterogeneously lived, *what differences do variances in bodies, embodied experiences, and worlds make?*

Multiple Realizability

How embodied mind theories in naturalized philosophy come out on the question of epistemic multiplicity is a complicated question. On the one hand, some of these theories conceptualize both cognitive processes and knowing agents as radically relational. For example, in arguing that cognition occurs without representation—that is, without an internalized cache of symbols that represent the world—Varela et al. (1991) posit that the entanglement of body-minds with environments opens up transient "micro-worlds" that are temporally and spatially specific. The argument is not only that perception is generated relationally, but that body-minds and worlds are too. For Varela, the breakdown of cognitive boundaries suggested a mind/body with no self, but rather a "multiplicity of micro-identities" (Protevi 2013), which are historicized through the traces left by previous cognitive events, or couplings of body-minds and worlds. This multiplicity is also possible in theories of extended cognition. For example, cognitive economies can be conceptualized as fixed or durable, but they can also be softly assembled, situational, and malleable. On the other hand, the effort to explain cognition at the phylogenetic level can have the effect of obliterating such onto-epistemic multiplicity. One can see this in the debate over *multiple realizability*, or whether the same mental state can be achieved through variant processes.

In a critical review of the literature, Clark (2008a) categorizes embodied mind theories in one of two types. In the first type, what he calls the *special contribution* thesis, bodily difference would, in principle, yield cognitive difference. According to this thesis, the specificity of sensorimotor dynamics fixes, "with extreme sensitivity, the nature of our perceptual experience" (52). In the embodied mind theory of metaphor, for example, the properties of the human body shape the content of concepts in a predictable manner. According to Mark Johnson and Tim Rohrer, for example, "Thousands of times each day we see, manipulate, and move into and out of containers, so containment is one of the most fundamental patterns of our experience. Because we have two legs and stand up within a gravitational field, we experience verticality and up-down orientation. Because the qualities (e.g., redness, softness, coolness, agitation, sharpness) of our experience vary continuously in intensity, there is a scalar vector in our world . . . . We are subject to forces that move us, change our bodily states, and constrain our actions, and all of these forces have characteristic patterns and qualities" (2007, 32). What can be known about the world depends on the biophysiology of the cognizer as well as the environmental circumstances that are more or less characteristic for all of us.

Theories of enactive perception also suggest that perceptual information is strongly shaped by a perceiver's motor activity (Mossio and Taraborelli 2008). That is, the meaning of stimuli is not a priori and neutral with respect to the body that receives it. Instead, meaning comes through the local relevance of stimuli for the bodily systems that are at work in the world. This view asserts, according to Clark, "a kind of principled body-centrism, according to which the presence of humanlike minds depends quite directly upon the possession of a humanlike body" (2008b, 43; see also Glannon 2009). Or, as Noe puts it, "If perception is in part constituted by our possession and exercise of bodily skills . . . then it may also depend on our possession of the sorts of bodies that can encompass those skills, for only a creature with such a body could have those skills. To perceive like us, it follows, you must have a body like ours" (2004, 35, cited in Clark 2008a, 41). In this sense, the special contribution thesis can be thought of as a humanist standpoint theory. The human body's properties are assumed to be both necessary and universal for cognition, and so the issue of cognitive difference (between humans) is foreclosed.

Unstated in his description of the special contribution thesis—for Clark is after something else—is that it necessitates drawing lines around what "possessing a human-like body" means. Do such bodies, for example, have two arms, two hands, two legs, and two feet? Are any of these vestigial, paralyzed, or amputated? Do bodies stand, bend, and walk, and all in the same way, and with the same effort or speed? Do they have equivalent haptic, optical, and auditory capacities? Do they have similar motor capacities? Do they require the same prostheses? In her critique of embodied mind theory, Emily Martin notes the pathologizing implications for disabled people by following the theory to its logical conclusion. If neural structures are generated through actions in the world that require universal capacities, Martin asks, "will abnormal individuals, for example those who are not able to 'go about the world constantly moving' . . . be unable to form the same cognitive structures as normal people, and hence be unable to participate in reason?" (2000, 572). Noe defends enactivism on this point by showing how various disabilities, for the most part, do not negatively affect perception. While "the enactive view requires that perceivers possess a range of pertinent sensorimotor skills," he reassures readers that even quadriplegics have such skills (Noe 2004, 12). For example, they can move using a wheelchair, and can understand the relation between movement and sensory stimulation. "Paralyzed people can't do as much as people who are not paralyzed, but they can do a great deal," he argues, and thus they have the necessary resources for enactive perception (12).

The parsing of the sorts of disabilities that do and do not lead to cognitive impairment is motivated by the aim of achieving a normative, singular, universalized model of cognition. In *Disability Bioethics* (2008), Jackie Leach Scully takes a very different tack. Scully finds it plausible not only that, as embodied realists and enactivists argue, "cognition is mediated through sensorimotor pathways laid down by the body interacting with the environment," but also that "this happens *differently* when anomalous interactions are involved" (103). Bodily difference yields cognitive difference. Like feminist epistemologists, Scully sees epistemic difference as a basis for critique of the normative and universalizing assumptions of philosophy. Taking a disability twist on embodied metaphor theory, she argues that common concepts reflect dominant morphologies, sensory experiences, and perceptive capacities, and are thus exclusionary. For example, concepts rooted in majority experiences of bodily boundaries, such as those related to autonomy and self-reliance, may be less resonant for people whose embodiment is rarely or never autonomous, such as persons who are highly dependent on caregivers for bodily maintenance. Metaphors rooted in verticality (e.g., "standpoint") are similarly shallow for those who spend their lives prone. Those dependent on sight ("blind spot") make little sense for those who hear the world but do not see it. People with variant perceptual and motor experiences do, quite literally, "see" and "feel" the world differently. Recalling Prinz's (2005) emotional theory of moral reasoning, Scully also argues that embodied appraisals vary, inasmuch as different bodies learn to expect different patterns of treatment from the world. A sense of outrage at bodily violation or intrusion, for example, may reflect experiences of privacy and bodily integrity that are not universally shared among people with disabilities who rely on caregivers for bodily maintenance.[15] Her point is that bioethics, which concerns itself with morality in relation to medicine, must take into account the ways in which moral knowledge is shaped by embodied experiences, which often are not shared by those who are the target of bioethical considerations.

Scully insists that neither physical nor epistemic universality can be assumed. Not only do body-subjects diversely interact with environments, but they do so with variable outcomes. Yet her argument also highlights the difficulties of conceptualizing physical embodiment as a source of epistemic difference. It raises the prospect of a physicalist standpoint theory of disability, or the idea that physical variation affords a distinctive vantage point on the world.[16] Recognizing the risks of this position, Scully takes care to disavow an essentialist conception of disability. She argues that if "adaptations of the environment are as distinctly formative of moral cognition as unusual morphologies, movements or perceptions themselves" (103), we cannot speak of something like a "disability mind" or a "disability brain." I argue further that bodily morphology and the environment are not separate elements with independent epistemic contributions. Rather, ability and disability—and other kinds of difference as well—can be seen in terms of the different ways body-minds couple or fit with various elements in the world. In other words, rather than seeing disability as an essential category of the body-subject, it must be seen (along with ability) as an experience in context, or "a particular aspect of world-making involved in material-discursive becoming" (Garland-Thomson 2011, 592).

While the special contribution thesis either dismisses the significance of physical differences between cognizers or implicitly pathologizes such differences as do exist, the second type of embodied mind theory Clark identifies, *extended functionalism*, allows for wide variation in the makeup of bodies and minds. Importantly, it does not regard differences as differentiating. In other words, it affirms the principle of multiple realizability, or the idea that the same process can in principle be achieved by multiple means. Clark's model places the human body in a larger system comprising multiple elements with which a body-subject might cognate. This view stresses cognition's environmental embedding and appeals "mainly, if not exclusively, to the computational role played by certain kinds of non-neural events and processes in online problem-solving" (2008a, 44). *Extended*, as I discussed previously, suggests that the network includes not only the brain and the rest of the body but elements of the outer world, whereas *functionalism* suggests "the nature of mental states is given by how they fit in a network of causes and effects" (Piccinini 2009, 513). Here, the body and environment are "merely additional elements in a wider computational, dynamical, representational nexus" (Clark 2008b, 49). A whole variety of configurations or assemblages could in principle achieve similar outcomes; what matters is the functionality of the overall cognitive economy. This thesis equalizes cognition across different bodies—including nonhuman ones—by allowing multiple configurations of brain/body/world to do equivalent work. It also opens up the body-subject to cyborg ontology, which is Clark's interest here.

Extended functionalism does not depend on an ideal body, or any body in particular; it would not distinguish between cognition achieved through a normative body with that achieved with a variant one. Neither would it discriminate prostheses from organic body parts, or differentiate between those who use prosthetic technologies as "disabled" body-subjects and those who use them for cognitive or physical enhancement. "What really matters" for Clark "is the complex reciprocal dance in which the brain tailors its activity to a technological and sociocultural environment, which—in concert with other brains—it simultaneously alters and amends" (2004b, 87). Clark's view offers a trans-human cyborg that overcomes whatever exclusions exclusively human embodiment might entail. Yet while extended functionalism does not require a normative body, the theory prefigures a functional fit within cognitive economies of brains/bodies/worlds. That is, it obscures variances and inequalities between different "fittings." *Both* types of embodied mind theory Clark identifies, then, try to account for common epistemic outcomes. Whether similar bodies create similar mental states (special contribution), or different bodies create similar mental states (extended functionalism), there is no attention to divergent mental states. There is no epistemic difference.[17] The question, then, is how to theorize the ontological multiplicity as Clark does in a way that allows for epistemic multiplicity or dissonance. Or, put another way, it is how to account for the difference that embodiment makes, without falling into a standpoint approach that essentializes difference as a property of the body-subject.

Disability as Assemblage

While philosophers of the embodied mind generally use disability as a negative case to explain "normal" and universal cognition—for example, asking whether quadriplegia interferes with "normal" cognition—disability studies puts bodily variance at the center of its analysis. From the perspective of many scholars in disability studies, neither ability nor disability is a fixed condition or vantage point. Disability "signals that the body cannot be universalized" (Garland-Thomson 2001, 2). Bodies regularly run afoul of neat categorization, and disability demands "a reckoning with the messiness of bodily variety, with literal individuation run amok" (2). While insisting on the capacities of disabled bodies is an important task in disability studies, equally urgent is "deconstructing the presumed, taken-for-granted capacities-enabled status of abled-bodies" (Puar 2009, 166). The ideal body has to be recognized as such; it is not a real, living, enacting, and perceiving body, but a construct that depends on the repression of disability, aging, and vulnerability (L. Davis 1995). All bodies are vulnerable (Turner 2006); further, the "instability of the disability body is but an extreme instance of the instability of all bodies" (Price and Shildrick 2002, 72).

Disability scholars commonly define disability as the effect of social forces that privilege some types of bodies and disadvantage others. The social constructionist model of disability argues that while impairment has physical aspects, what constitutes disability depends on the social organization of society and the particular capacities it values (Wendell 1989). The built environment, in particular, enforces social preferences in ways that generate debilities for those with variant bodies. Disability is also constituted by representations of particular attributes as unfit and is linked to other oppressions related to race, gender, sexuality, and citizenship (Baynton 2013). Some disability scholars have also made sense of disability as a collective identity; despite the wide range of experiences gathered under the umbrella of disability, those deemed disabled, they argue, have overlapping experiences negotiating the world (for discussion see Price and Schildrick 1998; Samuels 2003). Yet it may be in disability studies where the materiality of difference is articulated most persuasively, often in reference to lived bodily realities that veer far from the normative ideal. Bodies are physically different, and in some cases impaired, and social constructionism does not fully capture this (Baril forthcoming; Iwakuma 2006; Shakespeare 2013; Siebers 2001).

Increasing interest in the materiality of bodily variation reflects a turn in disability studies from a focus on representation to that of fleshly corporeality. For example, rather than seeing the body in the first instance as a site of cultural inscription, Tobin Siebers has insisted that the body is, "first and foremost, a biological agent teeming with vital and often chaotic forces" (2001, 749). Siebers advocates a "new realist" approach, one that takes bodily variation not only as real but also as consequential. This approach, like feminist neo-materialisms, adopts many of the critical insights of social constructionism while recognizing the material, biological body as agentic. Seibers argues for a theory of complex embodiment, where the physical body's significance is indebted to social meanings, but not wholly; rather, it also has its own capacities to shape experience. One of Siebers's key examples is the body in pain; in contrast to psychoanalytic and poststructural treatments of pain, Siebers insists that pain is not reducible to symbolic meaning, nor can it be understood as a kind of resistance to social inscription. For Siebers, the body "is not inert matter subject to easy manipulation by social representations. The body is alive, which means that it is as capable of influencing and transforming social languages as they are capable of influencing and transforming it" (751). But this does not necessarily mean that disability inheres in individual bodies. Assemblage theories of disability, like Rosemarie Garland-Thomson's (2011) idea of "misfitting," articulate ability and disability in terms of differential couplings of bodily capacities and built worlds.

If neurocognitive embodied mind theories see the environmental context as a collection of stimuli, conditions, and physical objects that are essentially similar for all human body-minds, or that vary significantly only in evolutionary terms, Garland-Thomson addresses the ways in which the environment is more and less functional for different body-subjects. She defines disability in terms of how well and easily a body-subject engages with the built world, a measure that shifts depending on the circumstances. For her, "the materiality that matters . . . involves the encounter between bodies with *particular* shapes and capabilities and the *particular* shape and structure of the world" (594, emphasis mine). Mis/fitting, then, is not a fixed condition or standpoint, nor a social construction, but an encounter: "Fitting and misfitting denote an encounter in which two things come together in either harmony or disjunction. When the shape and substance of these two things correspond in their union, they fit. A misfit, conversely, describes an incongruent relationship between two things: a square peg in a round hole. The problem with a misfit, then, inheres not in either of the two things but rather in their juxtaposition, the awkward attempt to fit them together" (592–93).

She gives examples of encounters of body-subjects and the built environment to suggest how misfitting is socially organized and context-dependent, while also being material. "One citizen walks into a voting booth; another rolls across a curb cut; yet another bumps her wheels against a stair; someone passes fingers across the brailed elevator button; somebody else waits with a white cane before a voiceless ATM machine; some other blind user retrieves messages with a screen reader. Each meeting between subject and environment will be a fit or misfit depending on the choreography that plays out" (595). Disability, then, inheres not in an individual body, nor in its representation, but rather in a relation that is temporally and spatially specific—a particular coupling of mind/body and world. Mis/fittings are embodied events, assemblages of body-subjects and worlds whose mis/fits are context- and interaction-dependent. How well an assemblage "fits" is not determined only by biological properties, nor circumscribed by identities; rather, it depends on many factors, including the features of the built environment, social norms, and particular aims, as well as the prostheses available in each encounter (Siebers 2008). The concept of "mis/fitting" draws attention not only to the body-subject's embeddedness in the world, and to the extended character of human action, but also to variances in that embeddedness and extensionality.

Body-subjects and worlds can both fit and misfit, and they can do so at the same time; that is, they can fit queerly. In some of Garland-Thomson's examples, misfittings have aborted intentionality. One gets to the stairs in a wheelchair and stops; one waits blindly by the ATM. Siebers (2001) highlights how body-subject and world can assemble in anomalous ways without being dysfunctional. For example, "Blind hands envision the faces of old acquaintances. Deaf eyes listen to public television. Tongues touch-type letters home to Mom and Dad. Feet wash the breakfast dishes. Mouths sign autographs" (737). Such assemblages could in theory be offered as evidence for extended functionalism; they show that the same task can be achieved by multiple means and with different resources. The dishes are washed; the face is remembered; letters are typed. With or without commonly expected bodily capacities, the cognitive and practical work gets done. Differences "need not be a problem" (Scully 2008, 16).

The ability to recognize this may be a significant payoff of extended functionalism. Yet Clark's approach masks the ways in which such assemblages can be more and less costly, can be more and less facilitated and encouraged by the social affordances of the world, and can generate novel outcomes. As Siebers complains of cyborg theory in general, "Prostheses always increase the cyborg's abilities; they are a source only of new powers, never of problems. The cyborg is always more than human—and never risks to be seen as subhuman. To put it simply, the cyborg is not disabled" (2008, 63). It is useful to recall that, for Haraway, hybridity is not to be celebrated or scorned; rather, it contains both risks and benefits, and these must be gauged in particular circumstances and contexts. Thus, the task is to ask, as Haraway does, "for whom and how [do] these hybrids work?" (1997, 280n1, cited in Sullivan 2001).[18] To put it another way, if extended minds are made possible by cognitive economies, how might these economies enact differences, even inequalities or privileges?

Embodied Multiplicity

Feminist epistemologies have identified the body as a vantage point of knowledge in order to challenge masculinist, universalizing rationalism and to address epistemic difference and multiplicity. The neurocognitive theories of embodied mind discussed above similarly challenge rationalism by demonstrating the inextricability of the physical body and mind. They also contest neuroreductionism by refusing to limit the mind to the brain. Their resonance with feminist epistemologies is clear in these respects. However, some of these theories tacitly or explicitly assume a universal body or a normative fit between body and world. Embodied realism uses a conservative lens of biological constraint to assess not only what epistemologically exists but also what is epistemologically possible. Its assumption of the commonality of the human body and its epistemic relation to the environment potentially leads to the re-instantiation of universal knowledge, the embodied counterpart to universal Reason. Extended functionalism, by contrast, does not aim to circumscribe the content of mind, and it acknowledges a wide range of differences in body-minds to include human and nonhuman elements; it echoes Haraway's themes of challenging human exceptionalism and opening up the body-subject to posthuman variation. But it presumes generic cognitive functionality, and it is unreflectively optimistic about cyborgian experience. In other words, it does not regard difference as differentiat*ing*. Thus a primary insight of feminist theorizing, that differences between knowers and contexts and worlds make a difference, is either undertheorized or undermined in this work.

Putting these literatures in conversation with disability studies highlights the need to address variations in bodies and embodiments, as well as to account for epistemic multiplicity and dissonance. Read together, the three literatures also offer lessons for exploring the differences embodiment makes. I want to summarize these as follows. First, models of embodied mind that shift the focus from the properties of the knower to the context of knowing enable an understanding of epistemic multiplicity that avoids fixing the subject. This does not mean that the properties of body-minds are irrelevant. But because the contexts in which they are embedded are temporally and spatially specific—that is, they are *events*—there are always elements of novelty as well as predictability. Embodied realism relies on discerning what the body can do, and feminist standpoint theory relies on discerning what subjects can know. Yet one could better argue that "Meaningful differences in knowledge and understanding are not features of knowers or their epistemic location, but patterns in the world that show themselves differently in different contexts" (Rouse 2009, 205).

This means that, second, the properties of body-minds are not fixed in advance with respect to cognitive tasks. Theories that pose context-dependent, relational roles for the various elements of a "cognitive economy," whether these elements are the positions of body-subjects or the arrangements of synaptic connections, can account for greater dynamism in cognition and mind. Instead of understanding body-minds exclusively in terms of elements that have properties prefigured by evolution, such as conserved neural patterns, or by social structures such as intersectional identities, they can be considered assemblages whose very assembling conditions the capacities they exercise. Assemblage theories in feminism and neurophilosophy, as well as in disability studies, call into question the predeterminacy of bodies and subjectivities, which seem instead to gain their boundaries in and through experience.

Third, theories that recognize the fluidity of boundaries between body-subjects and the world expand the complexity and malleability of mind beyond the individual, the human, and the organic, rendering exclusively biological or social frameworks insufficient. Such theories cannot assume that bodily difference makes no epistemic difference, or that all assemblages of mind/body/world are always equally functional, or that binary measures capture their epistemic outcomes. Couplings of body-minds and worlds can be more and less normative, more and less queer. At the same time, perspectives that emphasize malleability, fluidity, and boundlessness of mind still need to grasp how experience leaves its trace on body-minds; in other words, they need to acknowledge how experience can, over time, give us a sense of ourselves as subjects (Blackman 2012; Protevi 2009). Otherwise, they lead to an understanding of persons and assemblages as endlessly flexible, infinite *becomings* that are unaffected by experience, including power relations. The critical task, then, is to explore in what contexts and to what degrees the boundaries of body-minds shift, and precisely how and in what contexts these shifts may be variously enabling and constraining.

As my discussion in chapter 1 demonstrates, the recognition of multiple bodies, embodiments, and styles of cognition is not inherently unproblematic. Difference can be used to designate neural phenotypes, reify or essentialize categories of the subject, and contribute to neurogovernmentality. However, if the presumed universality of the body is an impediment to recognizing the varieties of human experience—and, importantly, the social inequalities woven through those variations—the presumed universality of embodied cognition operates in much the same way. I expand this argument in chapter 3, where I describe how embodied cognition is being grounded in neuroscientific research on mirror neurons. This literature addresses the body's role in intersubjectivity, but if embodied minds can mis/fit in the world, so too can bodies and minds encountering each other.

CHAPTER 3

I Feel Your Pain

A macaque monkey whose brain is wired up for single-cell recording sits in an Italian neuroscience lab on a summer day in the 1990s, much as she does every day, when a graduate student researcher walks in with an ice cream cone. Even though the monkey is sitting still (it is the graduate student who is moving with the precision required to eat an ice cream cone), the machine makes a noise to indicate neural activity in area F5 of the monkey's ventral premotor cortex. This is the brain area thought to be responsible for bodily movements related to grasping and interacting with objects. But when the machine makes its whir to indicate that cells have fired, the monkey has no ice cream cone; she is merely observing the human eating one. This turns out to be a very big moment not only in the Parma lab but also in the emerging field of social neuroscience. The detection of neural activity in this location in the brain, under these circumstances, led to the identification of mirror neurons, or neurons that are thought to fire both when an individual makes a motor action and when she sees another performing the same action. Some researchers now claim that in humans these special cells—which have reportedly also been identified in brain regions associated with facial recognition and pain processing—allow people to automatically grasp others' perspectives. That is, they allow "mind reading," or understanding another's intentions, and empathy, or feeling what she feels.

The enthusiasm about mirror neurons, and their divisiveness, is difficult to exaggerate. *Scientific American* describes a "mirror neuron revolution" that is utterly transforming our understanding of human social interaction (Lehrer 2008). The *New York Times* announced a decade ago that their discovery is "shifting the understanding of culture, empathy, philosophy, language, imitation, autism and psychotherapy" (Blakeslee 2006). V. S. Ramachandran calls them the "neurons that shaped civilization" (2011, 117). Marco Iacoboni (2008, 2009) suggests they may be the biological foundation for morality. Amy Coplan and Peter Goldie (2011) claim "it would be difficult to overstate the importance of mirror neurons, not only for the study of what we call low-level empathy but for our understanding of mental life more generally" (xxx). One can find references to them in not only philosophy and psychology but also theater and performance studies, aesthetics, literary criticism, art history, musicology, cultural studies, sociology, and more. In social theories of the body and affect, mirror neurons are used as evidence for the importance of feeling, embodiment, and intercorporeality over disembodied thinking, cultural inscription, and discourse.

Critics protest the expansiveness of such claims, and also raise methodological disagreements, disputes over their function, and even skepticism over whether they exist in humans at all (for discussion see Caramazza et al. 2014; Hickok 2014, Rose and Abi-Rached 2013).[1] Research that is continuing apace will be used to adjudicate these debates, for or against the existence of mirror neurons systems in humans, and for or against their relevance for understanding empathy, theory of mind, and other precognitive and cognitive functions. How mirror neurons will look in a decade's time is difficult to gauge, but I am interested in them as phenomena that demonstrate how the body and embodiment can be called into play in the biosocial brain.

Mirror neurons give the biological, fleshly body an important set of roles. The basic idea of the dominant model of "mirroring" is that the brain generates a grasp of the other, not with language and thinking but via simulated action and feeling. Intersubjective or social understanding becomes a bodily, prelinguistic activity instead of an intellectual or mentalist one. And perception and action are coupled, so that perceiving and understanding the other is tied to one's own capacity for moving and acting in the world. While the use of neuroscience to "authorize" such ideas is controversial (Blackman 2012; Hemmings 2005; Leys 2011, 2012), the notion of felt, interpersonally attuned, active perception and intercorporeality has a wide appeal in social thought. Scholars have argued that mirroring resonates with phenomenological, feminist, and neo-materialist perspectives on the significance of felt, embodied experience over abstract and disembodied cognition (Colebrook 2008; Krause 2010; Ravven 2003). Mirror neurons ground the affective view of communication as passing between bodies without need of representation (Connolly 2010, 2011). Mirroring suggests a mechanism by which individual bodies are innately social, and mirror neurons have been used to challenge the poststructural view of culture as primarily discursive (D. Franks 2010; Lizardo 2007).

Yet like other neuroscientific phenomena described in this book, the scientific articulation of mirror neurons does more than affirm the significance of the material body for sociality. It also defines what *sort* of body matters, what embodiment means, and what the social is. Most social theorists have paid relatively little attention to the particulars of how mirror neurons are theorized and enacted in the neurosciences. They often neglect the details of how mirror neuron researchers themselves adopt models from other disciplines, notably philosophy of mind and psychology, to make sense of their findings. To be sure, this leads to the reification of neuroscientific claims that in their original context often are considered to be still tentative, and the treatment of neuroscientific research as strictly empirical rather than also theoretical. More problematically, it also belies the multiplicity of neuroscientific practices, including the different neurons, brains, bodies, and embodiments they envision and enact.

In this chapter I unpack the dominant model of mirror neurons, which sees them functioning as *embodied simulation*. Despite claims to the contrary, the model does not inevitably lend itself to a view of the brain as dynamically social, of the body-mind as richly relational, or of embodiment as situated. Rather, it commonly brings forth mirror neurons as insular, atomistic entities shaped by evolution or fixed early in life. Its account of mirroring processes can seem generic, universal, and highly normative rather than ontogenetically specific and multiple. And its epistemic claims, which draw from particular conceptions of theory of mind and embodiment, are questionable. I join other critics concerned about the proposed universality of mirroring and its treatment as a basis for empathy and intersubjectivity, which (I argue) presents a skewed, optimistic picture of social relations and forecloses more politically astute assessments of somatic sociality.

However, other enactments of mirror neurons are possible. Just as DNA has been treated as both a "master molecule" that dictates traits and, alternatively, in epigenetics, as a "flexible form of material agency" emerging through the reciprocity of organisms and their environments (Weasel forthcoming, 2016), mirror neurons can appear in multiple and contradictory forms. Here I address these multiplicities and the prospects for thinking of mirror neurons in more ontogenetically specific, dynamic, and situated terms. A better grasp of their entanglement with social complexities can help rethink the terms of intersubjectivity and intercorporeality, from a natural relation out of which sociality emerges to an event in which nature/culture is transformed. This view calls into question certain claims about the sociality of the brain—does it depend on the universality of bodies, for example? Is sociality itself a matter of recognition, or does it include misrecognition and conflict? My interest here is not to advance one theory of mirror neurons over another, but rather to highlight different ways that the social brain can be understood, and to underscore the stakes.

INTERLUDE: WALLET VS. GUN

All of the talk about intersubjective understanding and empathy to follow in this chapter presumes that humans do generally understand each other, or that at least we do quite regularly, and that understanding in some way leads to empathy. I'll not downplay these features of human sociality. (Nor will I insist, as many do, that they are exclusively the outcome of explicit, symbolic effort, which overcomes an essentially less social bodily nature.) But there are constant examples of empathic failure and misunderstanding in everyday life, which seem to entirely disappear in neuroscientific discussions of these subjects except when cognitive pathology is discussed. Misunderstanding and failures of empathy can be devastating, especially in contexts marked by social inequality and violence.

As I worked on this chapter, New York City observed a tragic anniversary: fifteen years since 1999, when a police shooting of an unarmed man devastated city residents and drew national and international ire. The victim was an immigrant from West Africa named Amadou Diallo. I remember that year well; I had moved to New York to take my first job after graduate school teaching in the public university system. Diallo was hardly the first casualty of police brutality in New York, but we—my students, colleagues, and the international media, too—were all talking about the shooting with disbelief because the facts of the case seemed unfathomable. Diallo was twenty-three years old. On a February evening, four white police officers shot forty-one bullets at him on the doorstep of his own apartment building. One of the most troubling details was that the officers mistook his wallet for a gun. The officers were acquitted in a jury trial, but for many people the case remains a painful example of police racism. It is also sometimes cited as an example of contagious shooting, where groups of officers fire their weapons, without thinking, after one of them discharges his or her weapon (Saletan 2006).

The event is only one instance of racialized violence between unarmed citizens and police, which had occurred before and has been repeated since— including in Ferguson, Missouri, just as Diallo's friends and family were marking the fifteenth anniversary of his death. But it still strikes me, and many others, as a particularly poignant and devastating example of misrecognition— one that raises questions about how racialization and other power relations inflect perceptions of others.[2] Diallo was shot not simply because his wallet resembled a gun, but because to the police officers who killed him, Diallo himself appeared to the officers to be someone who was about to use a gun. I cannot help but remember him and others who have shared a similar fate, in stark contrast to the discussion of theory of mind, action understanding, and empathy that follows.

Neurons That Mirror

Vittorio Gallese, Giacomo Rizzolatti, and other neuroscientists working in the Parma lab first identified mirror neurons in macaques using recordings of the electrical activity of individual cells. Single-cell recording is an invasive technique; it requires the insertion of tiny electrodes directly into brain tissue in order to measure the action potential (electrical firing) of individual neurons. To study mirror neurons in humans, the researchers looked to neuroimaging labs such as the one run by Marco Iacoboni at UCLA. Using imaging technologies such as fMRI, which are much less invasive but also less precise, mirror neuron researchers have reported finding "mirror" activity in humans, not only in the premotor cortex but also in areas of the brain thought to be associated with affective and interpersonal processes (for a review see Grafton 2009). They make two fundamental claims about mirror neurons: they generate action understanding, and they are the biological substrate of empathy.

First, mirror neurons are thought to register another's actions in one's own brain as if performing the action oneself (but without actually doing so); in other words, they are thought to *simulate* the action of the other (Gallese and Goldman 1998). But mirroring is not meaningless imitation; rather, researchers argue that the neurons also afford a grasp of another's intentions. The neurons seem to work with considerable specificity. For example, some of the mirror neurons (described as "broadly congruent" ones) that fire in the picking up an ice cream cone scenario described earlier also reportedly fire when a research subject sees someone perform something different but logically related or having the same goal as that action, such as putting a cone up to her mouth (Iacoboni 2009). Researchers have observed what they think is mirror neuron activation even when an action is only partially observed (for example, some of her arm motion is obscured by a screen) but when the goal of her action is apparent (the subject can see that the ice cream is within the other person's reach). Mirror neuron researchers see this as evidence of *action understanding*, whereby the subject registers the other's goal for action (Iacoboni et al. 2005). This intentional attunement is thought to be precognitive and preconscious.

A second claim is that mirror neurons provide a biological foundation of empathy (Bernhardt and Singer 2012; Gallese 2001, 2003, 2009, 2014; Iacoboni 2011; Keysers et al. 2010). This claim is drawn partly from fMRI experiments that have reported mirroring in response to facial expressions and other communicative actions, and in response to seeing evidence of others' bodily sensations (Keysers et al. 2010). For example, a series of pain experiments reported activation in the somatosensory cortices when research subjects were exposed to cues signaling that someone else is in pain, seeing the facial expressions of others indicating pain, and observing pain inflicted on the hands and feet of others (Singer et al. 2006). Another study found neural activation in response to seeing another person's tactile experience, including being touched or stroked on the leg, hand, and face (Blakemore et al. 2005). Researchers argue that this activation, like premotor mirroring, amounts to the creation of a vicarious bodily state. "With this mechanism we do not just 'see' or 'hear' an action or an emotion. Side by side with the sensory description of the observed social stimuli, internal representations of the state associated with these actions or emotions are evoked in the observer, 'as if' they were performing a similar action or experiencing a similar emotion" (Gallese et al. 2004, 400). Such neural patterns are thought to generate an experience in one's own body of what another is feeling or doing.

Because of the neurons' ostensible role in action understanding and empathy, proponents of a strong view of mirroring argue that it provides the neural basis of intersubjectivity (Gallese 2009, 2014; Gallese et al. 2004; Gallese and Lakoff 2005; Iacoboni 2009, 2011). But critics complain that the empirical evidence for links between mirror neurons and meaningful intersubjective understanding are weak (Caramazza et al. 2014; Green 2009; Heyes 2010a, 2010b; Hickok 2009, 2014; Jacob 2009; Leys 2011, 2012; Saxe 2009; Wahman 2008), and they worry about extreme reductionism. As Cecilia Heyes warns, "mirror neurons are at risk of being viewed as atoms—primitive entities whose very existence explains a range of cognitive and behavioural phenomena" (2010a, 789). The embodied simulation model of mirroring, which was adopted early on by Gallese and his collaborators and is by far the most influential account, has generated some of the most expansive (and reductionist) claims about mirror neurons. The model draws from a particular tradition in philosophy of mind and a version of embodied realism to depict mirroring as a set of automatic and possibly hardwired processes, which are responsible for a set of basic social functions and also play a role in more complex ones. I describe some of the details below; later I argue that Gallese naturalizes a conservative view of somatic sociality, one that ignores conflict and empathic failure in everyday life.

Mind Reading without Words

How do I know what another person intends to do? To hypothesize the functions of mirror neurons, Gallese and others draw from debates in philosophy and psychology over theory of mind or mind reading, the ability of one person to understand the perspective of another without explicit communication. In other words, how one can know what another person is feeling or intends to do? On the classical cognitive view, mind reading is a higher-order process of intentionalist cognition. It is executed through the mental application of behavioral principles or folk theories about how other people think and feel to third-person observation. (Thus this explanation is called "theory theory.") For example, I know that the man I see wants to go into that building because I know that people usually go up the stairs and open the door when they want to enter a building. Perhaps I also know that the building is an apartment building, it is around dinnertime, and people are generally headed home at this hour. I apply such general principles from folk psychology to make propositions about other people's mental states. These principles are drawn from concepts that are symbolic representations of the world, whose validity depends on the extent of correspondence to it. Thinking, then, can be understood as the computation of abstract symbols to achieve an attribute of another's mental state. As Gallese and George Lakoff complain, this view inherited from analytic philosophy "the propensity to analyze concepts on the basis of formal abstract models, totally unrelated to the life of the body and the brain regions governing the body's functioning in the world" (2005, 455). Mind reading is a mentalist activity independent of embodiment and situatedness.

The primary alternative to the classical cognitivist view has been simulation theory, which has served as the main model for interpreting mirror neuron research. Advanced by Alvin Goldman and others, simulation theory argues that we understand another's actions not through applying generic principles but rather by simulating or pretending how we might act in the same situation ourselves. I know Lucinda wants to eat the cupcake she is picking up because I imagine myself picking up that cupcake and wanting to eat it. My understanding of her desire is in some way dependent on my ability to imagine my own desire, as well as my ability to see her as me and as not me at the same time. In Goldman's version of this, "the first stage of the imaginative construction is creation of a set of initial states (in the self) antecedently thought to correspond to states of the target (but not the specific state the mind reader wishes to ascertain). This is 'putting one-self in the other's shoes.' The second stage consists of feeding these inputs into one of the mind's operating systems and letting it output a further state. Finally, the mind reader 'reads' or detects that output state and projects it onto the target, i.e., attributes it to the target" (Goldman and Shanton forthcoming, 12).

Goldman's simulation theory rejects general, abstract reasoning in favor of a capacity to imagine oneself, within a particular context, in the shoes of the other. While theory theory "depicts mind-reading as a thoroughly 'detached' theoretical activity," simulation theory locates our understanding of others within our own self-awareness (Gallese and Goldman 1998, 497). In this sense simulated mind reading is less abstract, more personal, and more located in one's own situation. Both theories, however, ultimately see mind reading as a kind of theorizing about another's mental state and favor representational, rationalist mental knowledge over affective and corporeal experience. Feminist psychologists Katherine Nelson and colleagues object that in both scenarios the mind remains a "disembodied, autonomous, individually owned information processing or representational device . . . a cognitive mechanism, providing representations that mirror the real world 'out there'" (2000, 67). In their view, both mind reading theories are informed by assumptions that are "antithetical to the principles of feminist epistemologies" (68). These principles, as discussed in chapter 2, underscore the embodied and situated character of knowledge.

Although not entirely answering such criticisms, the joining of simulation theory to mirror neuron research rectifies its disembodied character. In their embodied simulation theory, Gallese and Goldman argue that mirror neurons "underlie the process of 'mind reading,' or serve as precursors to such a process" through embodied simulation (1998, 495). For his part, Goldman (2009) later makes a distinction between "low-level" automatic processes accomplished by mirroring and the higher-level processes of pretense and imagination. Mirroring, he argues, involves experiencing another's only relatively "primitive" mental states, such as disgust, pain, and anger, which can be exogenously induced by a subject observing another's action. Higher-level simulation, he argues, is more multifaceted, utilizes information, appraises more complicated mental states, and is endogenously produced by the subject's imagination. But on the strong view of embodied simulation embraced by Gallese and other members of the Parma research group, mirroring offers an "immediate representation of the motor acts being formed by others," and there is "no need for a higher-order representation" (Rizzolatti et al. 2006, 589; see also Rizzolatti and Sinigaglia 2008). Embodied simulation theory shifts the task of simulation from being the work of an effortful mind to that of a brain effortlessly grasping knowledge of the other, as automatic as contagious yawning or laughing (Gallese 2001). Through mirroring, putting oneself in another's shoes is a biologically mandated mechanism occurring at the moment of perceiving the other's motor, sensory, and emotional actions and experiences. This is achievable because, Gallese argues, the neurons draw from the body's own capacities for motor action, its relations with objects in the world, and the fact that, he claims, it shares these with other conspecifics.

The framing of mirror neurons as a resolution to the debate on theory of mind forces a choice between propositions, which require language, and shared neural representations, which do not. It reinforces rather than resolves strict divisions between thoughtful and felt, higher-level and lower-level, automatic and effortful cognitive processes; some argue that it discourages empirical and theoretical investigations of how these processes are related to each other (Zaki and Oschner 2009). This separation appeals to social theorists who insist on the autonomy of affect from language. William Connolly, for example, reads mirror neurons as an account of "how cultural practice becomes encoded into the human sensorium even before a child acquires linguistic skill" (2011, 797). But to see mirroring as both wholly autonomous and at the same time socially meaningful requires a particular set of claims about the universality of bodies, embodiments, and affordances that need to be interrogated.

INTERLUDE: MISREADING MINDS

If the story told in the courts and the media is to be believed, the shooting of Amadou Diallo is a disastrous example of a theory of mind failure. According to the account of the trial reported in the New York Times *(Fritsch 2000), the officers said that they did not consider the situation from Diallo's perspective. They did not, in other words, put themselves in his shoes, at least consciously. Nonetheless, they attributed goals to him.*

While acknowledging that they had made a mistake, the officers said Mr. Diallo was largely to blame for his death. He did not respond to their commands to stop, they said, and did not keep his hands in sight. Instead he ran into the vestibule of his building and began digging in his pocket, they said, and then turned toward the officers with something in his right hand. They said they thought it was a gun and began shooting, setting off a chaotic hail of ricocheting bullets and muzzle flashes that made it seem as if they were in a firefight.

When Mr. Diallo finally slumped to the floor, his wallet fell out of his right hand. There had been no gun.

In his closing argument, Mr. Warner (the prosecutor) suggested that Mr. Diallo may simply have been reaching for his wallet to hand it over to what he thought was a gang of robbers. Or perhaps, Mr. Warner said, he was trying to show the officers his identification.

The Times reports that Officer Carroll, who had yelled that Diallo had a gun and was the first to shoot him, sobbed as he recounted in court how he held the young man's hand as he lay dying on his doorstep; Carroll said that he felt "destroyed" by his error.

The problem was not only how a wallet can be mistaken for a gun (which, being neither the same shape nor the same size, requires some explaining about perception). It was also how a series of bodily actions can be equated with one set of intentions rather than another. According to their testimony, the officers assessed Diallo's intentions not primarily by the details of the object in his hand, but rather from observing him perform a sequence of actions—running away from them, putting his hand into his pocket, digging around, and then turning toward them holding something. In short, they thought his wallet was a gun because they thought Diallo was about to use a gun. Why? Is it because his actions fit their assumptions about how armed criminals behave? Is it because they were already prepared to see him as such?

Although intellectualist accounts of social cognition are rightly criticized for being too abstract and disembodied, they have the benefit of being able to easily account for failures. Simply put, theories of mind are wrong when the propositions people use to assess others (such as racist ones) are wrong. It is difficult to say precisely what these would have been in this case, but one wrong proposition explicitly advanced by Carroll was that Diallo fit the description of a suspect who had committed rapes in the area a year before. Another would be that Diallo didn't belong in the neighborhood. Yet another might be that young black men should respond in a particular way to being approached and questioned by police, or that people who fear the police are criminals. It is clear that propositional errors were at work in this event. But perhaps the theory is too rationalist to fully explain all forty-one bullets.

The simulation model advanced by Goldman could also explain errors. Seeing someone as like me, putting myself in their shoes, is one way to understand what another intends to do, but it does not guarantee success. In fact this strategy could lead me badly astray if the other person would not behave in the same way I would. Critics of simulation theory have pointed out that simulation could get in the way of, rather than promote, intersubjective understanding. A projection of one's own experience onto another person denies the specificity of the other's world. It refuses to recognize difference, say, between the thinking of a city police officer, who expects to see hands, and those of a young immigrant who expects to have to show his ID. Another possibility is that I am unable to see someone as like me at all if, for example, I view them as wholly different or alien. (I am on the right side of the law, he is not; I am American, he is a foreigner; and so on.) I am unable to put myself in their shoes, so I'm unable to grasp their intentions. This latter possibility could account for an absence of theory of mind. However, it doesn't quite explain a wrong theory of mind; I would have to resort to propositional thinking if I couldn't simulate, and the error would again be attributable to my mistaken beliefs.

Perhaps the mistake was not exclusively propositional, in other words, not wholly a matter of failed logic, presupposition, or ideology. Perhaps it was also affective. It's not irrelevant that this event often is cited as an example of contagious shooting.[3] Contagious shooting is a term used to describe how when after hearing or seeing one of their colleagues fire, officers fire their own guns, not accidentally, but without explicitly deciding to do so. They do not weigh the situation, nor do they consciously decide to trust their colleagues' decision to fire and join in. Instead, their explanation is that firing itself is contagious. "It spreads like germs, like laughter, or fear. An officer fires, so his colleagues do, too," as the Times puts it (M. Wilson 2006). In contagious shooting the agentic act of firing a deadly weapon takes place in some liminal space between thought and bodily deed, a space that does not end at the boundary of an individual body but extends to other bodies. If that is the case, its modification or curtailment cannot be achieved through exclusively rational means.[4] Contagious shooting may or may not have been a factor in the astonishing number of bullets fired at Amadou Diallo.[5] But the suggestion underscores the possibility that perception and action can be tied together at a bodily, felt level that does not involve explicit theorizing or conscious awareness.

The neural version of simulation has the benefit of addressing the bodily, felt, and contagious realm of interaction; this is why affect theorists have embraced mirror neurons. However, embodied simulation theory makes even less room for error or conflict than the other two accounts of theory of mind. As I explain, embodied simulation works—generates a felt attunement of the *other—because (goes the argument) our bodies really are the same, and really do act in predictable ways. There is no explanation for "wrong" action understanding outside of pathology. Embodied simulation cannot account, for example, for the sort of fear that attaches to bodies through racialization. To grasp such wrongs, we have to understand not only how propositional beliefs can be racialized but also how racism and other forms of power and inequality work with bodily capacities and perceptions, generating mis-attunements and foreclosing the possibility of intersubjective knowledge.*

Shared Embodiments

Gallese emphasizes the significance of bodily action in accounting for mirroring. Mirror neurons are special, he argues, in part because they can integrate sensory and motor activity within the motor system, which is involved in bodily agency. "Far from being just another species of multimodal associative neurons in the brain, mirror neurons anchor the multimodal integration they operate to the neural mechanisms presiding over our pragmatic relation with the world of others" (Gallese 2009a, 522). Individuals use the same or overlapping neural circuits for processing their own embodied actions and for processing the observed experiences of others. Contrary to classical "top-down" models that see interpersonal understanding as based on cognitive mind reading, where one rationally theorizes about another person's intentions, on this view understanding the other moves from the bottom up, allowing no less than "a direct experiential grasp of the mind of others" (Gallese et al. 2004, 396). This work contests the idea that "the sole account of intersubjectivity consists in explicitly attributing to others propositional attitudes like beliefs and desires, mapped as symbolic representations . . . before and below mindreading is *intercorporeality* as the main source of knowledge we directly gather about others" (Gallese 2014, 4).

This claim is grounded in the experimental research mentioned above, but it also draws from established ideas about perception and motor schema. It assumes first that objects have affordances for us—that is, we see them through the lens of how we might act in relation to them (Gibson 1966)—and second, that motor intentionality involves typical series of actions. When I see a chair, I do not see a wooden sculpture but a place to sit.

(If I were a cat I might see a place to scratch.) But I do not have to use high-level thought to grasp this—perception of the chair contains the parameters for action.[6] Embodied simulation theory suggests that this mechanism also works when observing others relating to objects. When I see Amy with her back to the chair and bending her knees, I do not have to consciously think "she is going to sit down"; action understanding is achieved through the firing of cells that would also be involved in the actions of sitting myself. This includes actions that Amy hasn't taken yet but would be the next steps in a pattern of motor actions (see Iacoboni et al. 2005). Mirroring is predictive (it can garner information about what the action is aimed at accomplishing before its completion) because for Amy and me, intentional actions are embodied in the same way. That is, they use the same motor schema. I would not have the same response if I were observing a creature—say, a robot—that uses a different motor schema. Gallese explains: "When a given action is planned, its expected motor consequences are forecast. This means that when I am going to execute a given action I can predict its consequences. Through a process of 'motor equivalence' I can use this information also to predict the consequences of actions performed by others. This *implicit, automatic,* and *unconscious* process of motor simulation enables the observer to use his/her own resources to penetrate the world of the other without the need for *theorizing* about it . . ." (2001, 41).

This motor equivalence, made possible by the impartial affordances of the world and the predictability of motor patterns, is no less than the basis of sociality because it allows for recognition of the other as a conspecific: "Action is the 'a priori' principle enabling social bonds to be initially established. By an implicit process of *action simulation,* when I observe other acting individuals I can immediately recognize them as goal-directed agents like me, because the very same neural substrate is activated as when I myself am bound to achieve the same goal by acting" (42).

We grasp others' intentions because, in one sense, we already share those intentions (they reflect our own motor schema). We recognize the other as like us, not through interpretive effort but because we feel the resemblance between our bodies and theirs, between our relations with the world and theirs.

This view of mirror neurons leads to several strong claims.[7] First, while mirror neurons work without cognition, introspection, or mental effort, they are understood to provide substantial intersubjective knowledge. "By means of embodied simulation, when witnessing others' behavior, their intentions can be *directly grasped* without the need of representing them in propositional form" (Gallese et al. 2009, 105, emphasis mine). Mirror neurons enable the "*direct comprehension* of the actions of others" (110, emphasis mine). Second, the neural mirroring of emotions is more crucial than cognitive modes of transmitting emotional meanings. Mirroring is "*the* fundamental mechanism at the basis of the experiential understanding" of others' actions and emotions (Gallese et al. 2004, 396, emphasis mine). Further, mirroring "scaffolds the cognitive description, and, when the former mechanism is not present or malfunctioning, the latter provides only a pale, detached account of the emotions of others" (396). So when I see Andrew cry, mirroring gives me a connected, fundamental experience of his emotional state. Without the benefit of mirroring, my comprehension of his crying would be shallow and indifferent. Finally, in contrast to cognitive efforts, this basis of our understanding of others is free of the ambiguity or errors of symbolic interpretation. Mirroring is a "mechanism or prepackaged process that normally guarantees success (i.e., genuine matching or resemblance)" (Goldman 2009, 246). Because it is not muddied by interpretation, mirroring supports an essentially objective understanding of the intentions of others; meaning is essentially transferred rather than created. That is, "the sight of other (living) human or human-like bodies *deposits in one's brain*" motor plans, sensory responses, and basic intentions of the other (Gordon 2009, n.p., emphasis mine).

Assembled with single-cell recordings of primates' brains, fMRI studies of human brain activity, debates in analytic philosophy and psychology, and theories of perception, affordance, and motor schema, this enactment of mirror neurons offers an alternative to intellectualist accounts of social cognition that have been widely criticized as too Cartesian, abstract, and disembodied. It breaks down dichotomies between mind and body, perception and action. It roots intersubjective understanding in felt attunement rather than propositional effort. It locates the "other" in one's own embodiment and via motor schema brings the active, experiential body into the mind-brain. But the enactment of biological relationality it achieves can be insular, limiting, and normative. Not only does the strong view risk neuroreductionism (attributing too much to mirror neurons or mirror systems alone) and determinism (fixing mirroring processes and the precognitive and cognitive functions they are thought to support). It also treats our capacities for understanding one another's intentions and felt experiences as universal. It accounts for difference, misrecognition, and dissonance only through pathology, with attention to cognitive disorders such as autism. In this sense it is heedless of the complexities of lived experience, such as how race, class, gender, and other social forces organize bodily experiences, differentially affect our relations to spaces and objects, and infuse individuals' perceptions of and feelings for others.

Biological Relationality

Beginning from the neuron outward means measuring what are perceived to be fundamental aspects of intersubjectivity within an individual brain. Obviously this approach risks solipsism and biological reductionism. Other approaches to intersubjectivity (which have their own limitations) begin externally, with the mutual interaction of body-subjects. In feminist psychology, for example, theory of mind is a product of "collaborative construction" that is "built out of experiential, pragmatic knowledge acquired in an interpreted, social world" (Nelson et al. 2000, 76). The experiential and pragmatic knowledge described in the embodied simulation model is neither collaboratively constructed nor acquired through interpretation. It is, however, at least in theory, shared and social. This biological sociality, which does not require language and conscious effort, is precisely what appeals to affect theorists, body-focused sociologists, and other scholars interested in how subjectivity is "bypassed in favor of a direct linkage of the social and somatic" (Protevi 2009, xi). Mirror neurons are social without being cultural in either the intellectualist or poststructuralist sense; they suggest the movement of information and feeling without need of discourse. As Lisa Blackman notes of affect theory, however, this approach "is in danger of reinforcing the very neuro-reductionism it is at pains to avoid . . . [when it] retreats to the *singular* neurophysiological body in order to explain the transmission of affect *between* people" (2012, 76).

To be genuinely social, biological mechanisms must be affected in some way by social interaction, which always occurs within contexts. While they should not be seen as "inscribed" by culture, to be social as well as material they must be entangled with collective patterns of experience. This includes not only patterns that foster mutuality and empathy but also those that differentiate populations, generate inequalities, and foreclose mutual recognition. And their epistemic outcomes—the attunement, empathy, and recognition they purportedly afford—cannot be monolithic and universal, but must be modulated in some way through experience. In the absence of these complexities the intersubjectivity we get from mirror neurons is all epistemic harmony, universal empathy, and "domesticated affectivity" (Slaby n.d., 2), without conflict, dissonance, difference—at best, a caricature of sociality, and certainly not one that most affect theorists endorse.

An immediate impediment to a genuinely social analysis is the claim that mirroring is biologically fixed or hardwired as the result of evolutionarily adaptation. The early accounts of mirror neurons by the Parma researchers and their collaborators lean heavily on evolutionary perspectives, for example, when Gallese admonishes, "When trying to account for the cognitive abilities of human beings we tend to forget that these abilities were not modelled and put in place as such, but they are the result of a long evolutionary process" (2001, 40; see also Gallese and Lakoff 2005). Initially, mirror neurons and their associated functions appeared not only as innate and automatic, but also unskilled. Although they operate in an intersubjective context, and supposedly perform functions that support intersubjectivity, the neurons themselves are not transformed by, but are already coded to respond to, the actions of the other. A later hypothesis is that mirror neuron systems are developmentally plastic. Infants are born with a rudimentary mirror neuron system, which is "flexibly modulated by motor experience and gradually enriched by visuomotor learning" in early life (Gallese et al. 2009, 106). Iacoboni (2008) suggests that while most mirror neurons are present at birth, some may develop afterward. Further, mirror neurons are "partially coded by experience" and can acquire new properties later on through learning (42). However, the need for a broader recognition of plasticity is acknowledged by Gallese, who claims that the more we learn about mirror neurons, the more we have to see them as situated in the personal histories of mirroring subjects. "By internalizing specific patterns of interpersonal relations we develop our own characteristic attitude toward others and toward how we internally live and experience these relations. It can be hypothesized that our personal identity is—at least partly—the outcome of how our embodied simulation of others develops and takes shape" (Gallese 2009, 531; see also Gallese 2014).

Nonetheless, Cecilia Heyes (2010b) complains that in most accounts of mirror neurons, "experience plays a relatively minor role in their development" (576). What is needed, she argues, is an alternative account to evolutionary adaptation to explain the existence of mirror neurons. Her "associative hypothesis" is that mirror neurons are created in individual brains through the experience of perceiving and performing comparable actions. "The individual starts life with visual neurons that respond to action observation, and a distinct set of motor neurons that discharge during action execution. Some of the motor neurons become mirror neurons if the individual gets experience in which observation and execution of similar actions are correlated—when they occur relatively close together in time, and one predicts the other" (2010a, 789–90). Experience, then, does not simply develop mirroring capacities, but produces mirror neurons in the first place (2010a). She points to fMRI studies of pianists, who show more mirror neuron activity than nonmusicians when watching a piano performance, and of dancers, who appear to have greater responses to watching other dancers. Her experiments with Caroline Catmur and Vincent Walsch (2007) enact mirroring through training body-subjects to move their hands and feet in particular patterns and retraining them to reverse mirror patterns. This flexibility suggests that mirror neurons are not precoded and mechanistically triggered but can learn and unlearn or recode with new experiences.[8] The associative hypothesis, as Cook et al. argue, necessarily "shifts the balance of explanatory power from MNs themselves to the environments in which they develop" (2014, 191–92).

Lessening the grip of evolution and making room for experience to shape mirroring is important, but it does not go far enough to render the mirror neuron account of intersubjectivity adequately social. Also required is a sense of how different patterns of experience shape different bodies' relations to other bodies and objects in the world. The embodied simulation model assumes that "human beings are basically similar to one another, with a limited range of variations" (Keen 2006, 212). Gallese (2001) argues that humans experience a *shared manifold* composed of a shared functional relation to the world, a shared phenomenological experience of it, and matching neural circuits in the form of mirror neurons. This not only "determines and constrains" (47) our understanding of others' intentions but allows us to "recognize [others] as similar to us" with "ease" (42). The idea that this recognition of the other as a conspecific, as Gallese claims, amounts to empathy is seriously questionable (Debes 2010; Saxe 2009; Slaby 2013; Wahman 2008). More fundamentally problematic is the insistence that our embodiments—our relation to objects, our phenomenological experiences, and our motor schema—are all so precisely equivalent to begin with.

Sociologists of the body describe embodiment as structured through social interactions and practices, and therefore as culturally differentiated. Pierre Bourdieu (1984, 1990) proposed the concept of *habitus* to address social reproduction (especially class differences) as effects of action and embodied practice rather than identity or consciousness. In his oft-cited example, a tennis player doesn't plan each stroke, but she acts with skill nonetheless, by using her "feel for the game" (Bourdieu 1990, 66). Similarly, people have an embodied feel for the social fields they inhabit, not as the result of conscious deliberation, but of a less reflective, more embodied absorption of local collective practices. Class differences involve more than merely economic and ideological differences—they are deeply embodied—because different fields produce distinctive habitus. As Omar Lizardo notes, "the most important practical competences constitutive of the class habitus are never the subject of explicit instructions, but are 'picked up' by the actor by virtue of being surrounded by other actors who display the same competences" (2007, 337). Loïc Wacquant (2004, 2015) demonstrates this ethnographically in his study of a boxing club in a poor neighborhood in Chicago. Wacquant describes how the social organization of the setting, the pacing of events, the collective physical work, the timing of movements, the social roles of various actors (coaches, boxers, and so on), generate intercorporeal interactions that are more telling than explicit linguistic communication. The culture of the boxing club as he sees it is not primarily discursive, but carnal.

Lizardo finds in mirror neurons, and in embodied simulation theory specifically, the empirical grounding for Bourdieu's and Wacquant's view of culture as physically enacted, rather than "downloaded" from a "collective object" (2007, 324). Following the embodied realist tradition, Lizardo argues that it is "the human body itself" that provides the shared conditions for cognition and therefore provides a nonlinguistic basis for the enactment of culture. But if bodily practices are inflected with and constrained by external structures that organize their temporal, motivational, normative, and spatial aspects, then so are embodied simulations. Lizardo writes, "any social setting that acts directly upon the body for a given collective will necessarily result in the sharing of similar 'practical presuppositions' about the world. The reason for this is that any ecological and/or social technologies that serve to modify the body (Wacquant 2004) will also result in the transmission and 'embodied simulation' (Gallese 2000) of other members of the group of similar bodily techniques, and thus the 'picking up' of the embodied concepts embedded in those patterns of practice" (343).

Lizardo is primarily concerned to address how the body "serves to highly delimit the natural diversity of possible logics or 'forms' of practice" (346). However, in the logic of Bourdieu and Wacquant, differences in social environments would create different outcomes. Boxing clubs and finishing schools would result in different practical competencies (and, potentially, different motor schema). Departing further from Lizardo's argument here, I think it also follows that such differences could also lead to embodied conflicts and dissonance. I argued in the previous chapter that a focus on embodiment requires that we reject, rather than adopt, generalized and universalizing epistemologies. As Bourdieu recognizes, the bodies we bring to interactions are not generic but historicized. What's more, embodiments are not all the same, even when we find them together in the same settings. Embodied social interactions are not best understood as isolated and seamless couplings of matched beings, but rather as complex assemblages of multiple actors and intersecting conditions.

In Gallese's account, embodiment seems to be not so much a variable, diverse situation as a stable and universalizing source of experience. Embodied simulation theory assumes, for example, that a brain should echo the neural firings of an observed person's brain in certain conditions, or that two different brains would generate the same intentional attunement of an observed other. Failure to echo neurally another's felt experience or intentions—in other words, requiring explicit effort to understand them—means that one's mirror neuron system is broken. Embodied simulation theorists have advanced this as an explanation for autism and are pointing to fMRI studies that have found diminished mirroring in the brains of people diagnosed on the autism spectrum (Gallese et al. 2009; Gallese et al. 2013; Iacoboni 2008; Vivanti and Rogers 2014; Williams et al. 2001, 2006). Based on the idea that theory of mind is simulational rather than propositional, they argue that social deficits symptomatic of autism are due to deficits in motor cognition, rather than an inability to theorize the perspective of another. This "broken mirror" theory of autism is controversial since autism seems to have a complex etiology and manifests in a wide variety of traits, and since mirror neurons are still poorly understood (Enticott et al. 2013; Hamilton et al. 2007; Leighton et al. 2008; Pineda 2008; Southgate and Hamilton 2008). The hypothesis also competes with other theories of autism. For example, two other neuroscientific research programs I describe in this book—on the sexed brain and on oxytocin and social bonds—have also proposed to be relevant to autism. My concern here is that there is little consideration of simulation failures outside of pathology. The hypothesis also problematically assumes that there are (as yet unspecified) mirroring processes that can be said to be *neurotypical*.

The dynamism of embodied simulation theory, to borrow Papoulias and Callard's phrasing, is potentially "arrested at one end by evolution" (2010, 47) and at the other by a readily normative and singular model of mirroring. The model does not exclude possible roles for learning and social context, but it fails to attend to how learning and context influence what are thought to be automatic neural processes. By design, the model also fails to acknowledge, outside of pathology, diversity in brains, persons, and situations that could lead to epistemic differences and intersubjective and intercorporeal conflict. It is not enough to propose that mirror neurons are plastic and situated; also required is a consideration of how this plasticity disrupts the homogeneity of embodiment across persons. In other words, it has to address what difference context and experience make for the bodies that are engaged in social interactions, and what differences the interactions themselves create.

Situated Neurons and Embodied Perception

Critics of embodied simulation theory reject the idea that mirroring can account for the complexity of either theory of mind or empathy. For example, each of these may require not merely simulating what another person does or feels but also sensing and grappling with "relevant differences in disposition, ability and character," hurdles that cannot be overcome without the use of more representational cognitive processes, if not actual communication as well (Nelson et al. 2000; Saxe 2009; Wahman 2008). In other words, theory of mind and empathy make multiple demands on body-subjects and therefore draw from multiple kinds of resources; they cannot be accomplished by mirror neurons alone. Jessica Wahman (2008) points out that physiological responses to others' actions are not necessarily a reliable basis for knowing their intentions. In fact they may frequently be responsible for misattributions of others' intentions. "The automatic firing of sensorimotor neurons may serve as a condition for the possibility of this achievement [of communicating meaning]; but by itself, such activity not only falls short of shared meaning: it can stand in the way of it as well" (178). Further, shared meanings must be accomplished jointly, through interpersonal activity. "By this reasoning, mirror neurons, or even a whole neural network that includes the limbic system and the insula, cannot achieve something like empathy or inter-subjectivity in the absence of language" (Pitts-Taylor 2012, 178).

One solution to this overreach within the mirroring literature is to allow that the neurons enact only low-level or basic cognitive functions, rather than more complex ones (Goldman 2009). For example, mirror neurons might allow me to grasp only basic emotions (pain, disgust, anger) but not more complicated ones (envy), or they might achieve the limited function of allowing me to recognize another as a conspecific, but not the broader one of generating empathy for her. Distinguishing between lower- and higher-level processes is a common strategy for explaining how singular biological mechanisms contribute to multifaceted psychic events. The distinction preserves the fixity and universality of those processes deemed basic, allowing only "higher-level" processes to be dynamic. In the mirror neuron literature this strategy leaves intact a precoded mechanism that provides scaffolding for, and is a constraint to, more complex social cognition, and it forecloses the question of how preconscious processes are themselves modulated through experience. To take seriously the idea that mirror neurons are plastic and situated is to address how they are entangled with social practices. The focus must shift from measuring how much these neurons on their own "mirror" reality to how they participate with other agents and capacities in generating intersubjective events.

Neuroscientific practices bring forth particular enactments of mirroring; these are often basic, autonomous, and isolated by design. As Zaki and Oschner (2009) point out, experimental studies often provoke what they deem to be mirroring in response to very simple stimuli (visual or auditory) within narrowly sketched scenarios. This is the case even though the sorts of social events they ultimately seek to explain are multimodal, involving a whole range of stimuli at once. Social scenes also have temporal complexities that are not typically reflected in lab experiments. Even if isolated motor actions have predictable schema, social scenes involve multiple actions that unfold dynamically over more than one time frame. Further, the perception of another's motor actions takes place in the presence of prior information (Zaki and Oschner 2009) and in the wake of previous experience.[9] Damasio and Meyer (2008), for example, postulate that the activation of sensorimotor neurons does not occur on its own but in combination with memory systems. In their view action understanding "is not created just by mirror-neuron sites, but also by the nearly simultaneous triggering of widespread memories throughout the brain" (168). The same might be true for empathic understanding at the level of "basic" empathy (Stueber 2012). The outcome of mirror neuron activation would thus be affected by the somatic history of the body-subject, the traces of experience felt in the present. To extend Prinz's (2004) argument, if mirroring plus memory amounts to something like an embodied appraisal of the other, it would be hard to say that such an appraisal is entirely devoid of symbolic meaning.

Mirror neurons often are treated atomistically to prove their independence from symbolic representation. When they are explored as agents that co-act alongside other processes, systems, and contexts, however, their full autonomy is called into question. To take one example, a series of experiments by Shirley Fecteau et al. (2010) aimed to measure the effects of symbolic priming on mirror-related motor activity—in other words, how context can confer meaning to a stimulus, and how this meaning affects the neural response in the premotor cortex. The researchers used transcranial magnetic stimulation (TMS), a technology that generates a pulse of electric current to stimulate activity in a specific area of the brain, on the left premotor cortex of fifty-two volunteers. Inducing brain activity in the left premotor cortex affects nerves in the right half of the body. By placing electrodes on the wrists and palms, Fecteau et al. calculated the neural responses (motor evoked potentials, or MEPs) to the pulses administered in the subjects' right hands. The broad aim of the experiment was to explore whether they could modulate the effects of the pulses (as measured by MEPs) by exposing research subjects to visual stimuli, and to discover the conditions under which different stimuli could affect those responses. In other words, could they modulate neural firing by adding meaningful context?

In the first experiment Fecteau et al. showed the subjects pictures of right and left hands, asking them to count fingers, and pictures of five dots that were placed in positions that sparsely mimicked the tips of fingers in right- and left-hand positions. In between they administered TMS pulses and measured the amplitude of the MEP in the subjects' own hands. They reported that the effect of the visual stimuli on the excitability of the motor cortex differed between the hands and the dots. When subjects were shown pictures of hands, researchers recorded a difference between seeing pictures of the right hand and the left. (More activation was recorded in the right wrist when seeing pictures of the right hand than pictures of the left hand.) The pictures of dots had no such effect between left and right. In the second experiment they wanted to see whether prior exposure to the hand stimuli could make a difference in how the dots modulated the firing of the premotor cortex. In other words, could they manipulate the conditions so that the premotor cortex would respond to the dots in the same way it responded to the hands? They showed one group of subjects two blocks of hands, then a block of dots, and compared this with a group who saw only dots. The first group of subjects (those who saw the hands first) showed greater excitability to the dots (between right and left orientation) than those who did not. Fecteau et al. believe this means that a dot can be neurally received as a representation of a hand in the context of prior information: "a dot in itself does not elicit a specific response but can become meaningful depending on what the subjects believe the dot motion represents" (175). A dot can be a hand, in other words, if I recall the hand when seeing the dot and make an inference. Fecteau et al. argue for "significant malleability in the way that the mirror neuron system codes observed actions and contributes to action understanding" (176). The dots/hand experiment suggests that mirroring can be neuroscientifically enacted as a differential and differentiating process rather than a generic or universalizing one.[10]

An atomistic picture of mirror neurons sees them as generating meanings about the other in concert with the impartial affordances of the world and the universal bodies and motor schema of conspecifics. The result is the transfer of objective information between bodies about subjective and phenomenological states, and thus the production of intersubjective knowledge as both pure intercorporeality and, somewhat ironically, as epistemically neutral. The forms of intersubjectivity this approach describes emerge out of events in which individual body-subjects have occasion to observe each other interacting with objects in the world. However, the intersubjective knowledge itself is not emergent, and certainly not divergent (unless pathological); rather, it is predictable, reflecting already drawn out motor schema and affordances that do not themselves seem to be changed in the interaction. Alternatively, mirror neuron systems can appear as part of a situated perception—inflected by learned, symbolic associations that may differ or differentiate. If mirroring is developed by associative learning, and through exposures to social interaction, it can be transformed, enhanced, or reversed through experience. This suggests that mirror neurons do not receive neutral meanings about the other. Rather, body-subjects and worlds bring particularities to present encounters of the other at multiple levels and are themselves transformed in such encounters. The epistemic outcome would be joint or collective (Slaby 2013) and emergent (Murphie 2010) and would necessarily vary across combinations of persons, conditions, and situations.[11]

Empathy, Alterity, Otherness

Do racialization and other forces make a difference in how people know each other? On Wheeler Avenue in the Bronx where he was killed, Diallo is memorialized on a huge mural that also depicts the Statue of Liberty and the words *American Dream*. This image underscores how generalized concepts of collective harmony (American Dream) obscure painful conflicts and differences. By juxtaposing the story of Diallo with the discussion of embodied simulation, I mean to make a similar gesture. The use of the neuro-biological body to define theory of mind and empathy as forms of simulation has, in the dominant model discussed here, the effect of normalizing intersubjectivity as a natural and reliable outcome of social encounters. Further, the use of theory of mind and empathy to define what mirror neurons do has the effect of limiting their functions entirely to prosocial roles. But embodiment can be understood as marked by inequality; affected by race, class, gender, and other patterns of social difference; and enmeshed in suffering and violence, as easily as it can be viewed as a common thread that unites. Embodiment is not exactly the same for everyone, and simulation cannot guarantee sociality or empathy. The potential for conflict, misunderstanding, and violence should not be set aside, nor acknowledged only as clinical pathology, but rather understood as part of embodied reality in contexts of persistent inequality.

If imagination is a "fundamentally embodied capacity of mind," Scully argues, "having/being a particular kind of body places real constraints on our capacities both to imagine ourselves otherwise and to imaginatively put ourselves in the place of others" (2008, 55). On Scully's view, bodily variance may disallow a shared manifold that guarantees understandings across boundaries of self and other. But it is not only physical dis/ability that affects phenomenological and functional experiences in the world. Social environments enable differing practical competencies, provide varying affordances, and require greater and lesser effort to "match" or fit with the surroundings. They invite (and disinvite) styles of bodily being, provide varying levels of safety and risk, and expose persons to different opportunities and constraints. Some bodies, for example, are subjected to daily surveillance and stop-and-frisk policing, whereas others are given free range to go about as they please. These differences surely complicate intersubjectivity. They probably mean that we cannot rely on simulation, whether propositional or neural, to do the work of knowing the other and of relating to them and feeling for them in nonviolent ways.

The need to recognize such differences is why social workers Karen Gerdes et al. qualify *social* empathy as "the ability to understand the circumstances of other people's living conditions in the context of broader educational, health, and socioeconomic structures and institutions" (2011, 85). Others describe *critical empathy*, where felt connections to another do not give way to "overidentification" (Weil 2010, 16), but rather allow that the other may be different from oneself. For intersubjective feelings *for* the other to be reliable—for them to afford recognition rather than erasure—they may require a sense of alterity, the irreducibly of another's experience to one's own (Levinas [1970] 1999, cited in A. Franks 2004). The recognition of irreducible otherness is no doubt risky, as when difference is sedimented and "finalized," or when it "reduces this person to some set of inherent properties to explain why he or she is a problem" (A. Franks 2004, 117).[12] Despite these risks, it may be necessary to acknowledge the limits of shared embodiments, shared visions, and shared knowledges in order to pay more serious attention to the failures and violences of intersubjectivity.

CHAPTER 4

Neurobiology and the Queerness of Kinship

PRELUDE

When the wedding photo of Lela McArthur and Stephanie Figarelle went viral, it was not because the couple looked great in their tuxedo and wedding dress (they did), or because the setting was so spectacular (it was). It was, rather, because the most popular conservative media outlet in the United States, meaning to illustrate nuptial bliss, used the image to accompany a column titled "To Be Happy, We Must Admit Men and Women Aren't 'Equal.'" Repeating a familiar premise, the column's author Suzanne Venker (2013) argues that the decline in marriage rates in the United States can be blamed on forty years of feminism, with its claim that "gender is a social construct." While the "cultural script" of gender equality has confused people into thinking that men and women are the same, Venker claims, we would do better to value the natural differences between them. When we ignore these differences, no one knows "who's supposed do what," leaving young people to battle the war of the sexes indefinitely and rendering marriage undesirable. I won't bother to argue with Venker. Rather, I want to recall the egg on the news organization's face when it became clear that McArthur and Figarelle were not a great illustration of traditional marriage (Breslaw 2013). In fact, they are a lesbian couple, the first same-sex couple to be legally married at the top of the Empire State Building. The much-discussed blunder highlights the article's exclusion of an ironic fact about marriage: Despite the overall decline in marriage rates, many people are actually clamoring to get married, so that their kinships, the enduring bonds they make with and feel for specific others, will be legally and socially recognized.

The misidentification of the happy couple as straight results from the heteronormative gaze, a way of looking that erases not only individual bodies and subjectivities that do not conform to binary genders but also interpersonal bonds that do not comply with heteronormativity. This gaze reflects the naturalization of kinship as emanating from reproduction and fixed by evolution. The participation of the biological sciences in articulating this understanding of kinship is well documented and has been the subject of considerable feminist and queer critique. For many critics, kinship, not unlike marriage, is a social construct, a projection of gender norms onto biology, and as such it must be either radically reformed or rejected. Kinship can also be understood as affective, however, potentially exceeding heteronormativity through feelings and ties that do not adhere to social norms and rules (Eng 2010). In this chapter I explore the treatment of the neurobiological body as a scene of affective kinship, a site for the generation and feeling of ties to specific others, in both humans and nonhuman animals. The neuroscientific enactment of such bonds is principally heteronormative, I argue here. Yet there are also queerer potentialities, including moments of misrecognition that reveal the complexity and multiplicity of affective ties in not only humans but also nonhuman animals.

Stories of Kinship

Kinship is a biocultural category; it can, for example, encompass both blood relations and socially arranged alliances, as well as the ways these constitute each other through the patterning of mating and reproduction. To put it another way, kinship is "a technology for producing the material and semiotic effect of natural relationship, of shared kind" (Haraway 1997, 53, cited in Norton and Zehner 2008, 117). It is also what Myra Hird calls the "strongest structure of exclusion" and inclusion (2004, 222). Kinship is a highly contested phenomenon because what counts as kinship is implicated in, and has immediate and material bearing on, almost every aspect of social life: on legal, political, religious, and economic institutions; on identities, relationships, and communities; on health, well-being, and longevity; on work and leisure; on citizenship; on patterns of social equality and inequality; and on and on. It even matters to time: What counts as kinship affects the meanings of the past, the comparative weight of the present, and the ways the future can be imagined (Halberstam 2005; Freeman 2010).

I want to begin by contrasting two very different ways of thinking about kinship. In the first story, articulated in contemporary rights movements, kinships are made through social structures imbued with power relations. This view has roots in structuralism and social constructionism. In her seminal essay from 1975, "The Traffic in Women," Gayle Rubin follows Lévi-Strauss and Engels in describing kinship as the effect of the cultural and economic organization of social life. Lévi-Strauss saw kinship as the result of universal cultural rules underlying patterns of marriage and reproduction, whereas for Engels it is tied to the organization of labor. Rubin's crucial insight is that through these mechanisms, kinship is also responsible for constructing gender. The primary target of Rubin's essay, which remains influential for its forceful articulation of the social construction of a "sex/gender system," is biological determinism. She ultimately defines kinship as "the imposition of cultural organization upon the facts of biological procreation" (G. Rubin 1975, 170–71, cited in E. Wilson 2010, 201). Crucially for Rubin, if culture rather than biology provides the rules for kinship, then kinship structures, and thus gender, are amenable to social and political transformation. (It is this sort of view that rankles conservatives.)

The second account, the biological tradition contested by Rubin, is rooted in Western conceptions of kinships as organized exclusively through blood ties and reproductive marriage. Unlike numerous examples described in cultural anthropology where kinship and blood are not faithfully coupled, the Western mode of determining kinship has long been focused on consanguinity and is now predominantly genetic. The biogenetic account is heteronormative, that is, seeing kinship as inherently heterosexual and reproductive. It assumes that males and females have bifurcated roles, which are universal and immutable. Social neuroscience is both extending and shifting this biological account. Social neuroscientists working on neurohormonal systems articulate kinship not in terms of immutable genes, but neural systems that generate feelings of attachment. Some social neuroscientists argue that humans and other mammals are able to experience kinship bonds through the involvement of neural systems linked to affect and memory. Of special interest is the role of the neurohormone oxytocin in the experience of attachment, defined as "the dispositions to extend care to others, to want to be with them, and to be distressed by separation" (Churchland 2011, 201). Ignoring the social, legal, and economic hierarchies of human kinship structures, social neuroscientists have emphasized kinship from the biological ground up: the felt ties upon which, as they see it, kinships depend.

Kinship is deeply embodied in the neurobiological story; it is built not on the rules of culture, but on the body's capacities for generating intercorporeal bonds. This focus on felt ties addresses in some way what can be called *affective kinship*. Rather than a mechanistic reflection of genetic rules, kinship becomes a matter of lived feeling that requires intercorporeal contact. This account potentially opens up the biological understanding of kinship as embodied and relational. As I describe in this chapter, however, much of this research is underpinned by the assumption of sexually dimorphic biologies and evolutionary imperatives. Researchers make heteronormative claims about sex, monogamy, and parenting that erase queer and nonprocreative ties and attachments.

If kinship is constituted by, in Judith Butler's phrasing, "relationships of various kinds which negotiate the reproduction of life and the demands of death . . . that emerge to address fundamental forms of human dependency, which may include birth, child-rearing, relations of emotional dependency and support, generational ties, illness, dying, and death (to name a few)" (2004, 103), there are now many visible patterns of kinship that challenge dominant norms. Social transformations in the past four decades, involving both new reproductive technologies and also the successes of the feminist and gay rights movements, have challenged the Western coupling of reproductive and kinship roles. The biogenetic kinship narrative has not disappeared, but new technologies and the social practices they enable underscore how kinship is, as Hird argues, *underdetermined*. In vitro fertilization (IVF) and other forms of assisted reproduction technologies, egg and sperm harvesting, surrogacy, and transnational adoption are widely used by straight, gay, cisgender and transgender people to reorganize traditional family patterns. Perhaps we are already in a future where "genetically referenced categorizations of kinship will hold less practical significance for many people" (Norton and Zehner 2008, 106). In this context biological stories about kinship emanating from male-female reproduction seem not simply outmoded, but reactionary and defensive.

Perhaps new forms of kinship underscore the sheer irrelevance of the biological body. Perhaps they prove something like the triumph of choice over nature in the private sphere, an idea captured in the title of Kath

Weston's (1991) ethnography, *Families We Choose*. If kinship is a social construction, it can be rescripted to enable new social, cultural, and legal arrangements that respect and foster alternative ideas of family, particularly those that do not depend on biological relatedness. Alternatively, poststructuralist theory teaches that such scripts are not the result of choice, but rather historicity and power relations; for this reason Butler (2004) proposes we might do better without kinship at all. For Weston's research subjects, kinship is a set of rules that can be rewritten, whereas for Butler it is rooted in discourse that is "always already heterosexual" (102). Either way, kinship is understood as wholly representational, not natural but cultural, not material but discursive. (The loss or denial of kinship, then, is primarily a loss of rights, recognition, or cultural legibility.)

I want to pursue a different line of thinking. Queer forms of kinship challenge the heteronormative order not because they overcome, or have nothing to do with, nature or biology. Rather, they demonstrate the reality of felt, affective bonds that do not follow heteronormative and reproductive patterns. They show, as David Eng puts it, that "the *feeling* of kinship belongs to everyone" (2010, 198). Eng's work describes affective kinships as deeply felt, embodied ties that are experienced in all sorts of kin relations that transcend biogenetic models of family, including queer and transnational ones. Are such bonds—whether "chosen" or not—embodied and entangled with biological being? Can the body's contributions to affective kinships be acknowledged outside of heteronormative assumptions? How might a feminist and queer critique of kinship recognize not only culture's capacity to be heteronormative but also nature's capacity to be queer? The problem is not simply to rehabilitate kinship as less heteronormative, but rather to *rethink what it is to be biologically related to another.*

Social neuroscientists offer accounts of kinship that are embodied, felt, biological, and intercorporeal, and also in many cases relentlessly heteronormative. Drawing on animal research, neuroendocrinology, evolutionary biology, and experimental psychology, among other fields, they trace what they see as the bonds of kinship, or "attachment," to their brain processes and structures, with special attention to the oxytocin and vasopressin systems. In this chapter I give an account of attachment research as it is being practiced in social neuroscience. I focus on the role of oxytocin in pair bonding, monogamy, and the theory of *brain maternalization*. This work brings together a set of scientific disciplines and research agendas, assumptions about evolution and gender roles, polypeptides and receptors, the bodies of human and nonhumans, and their relations to enact and measure attachments. While this research is being applied to explain reproductive kinship, the heteronormativity of this work is not inevitable; neurobiological systems can be understood otherwise.

Oxytocin and Social Neuroscience

In 1909 Henry Dale treated the uterus of a pregnant cat with an extract taken from the human pituitary gland, causing her to begin labor. He named the substance oxytocin (from ancient Greek, meaning "quick birth"). A year later, Ott and Scott described its role in milk-ejection in mammals. In 1953 oxytocin was the first polypeptide, or compound of multiple amino acids, to be sequenced and biochemically synthesized. Oxytocin and a similar polypeptide, arginine vasopressin, are now known to be hormones made in the brain by cells in the hypothalamus.[1] They are stored in the axon terminals (endings) of neurons that extend from the hypothalamus to the posterior pituitary gland, located at the base of the brain, where they are released into the bloodstream for peripheral circulation throughout the body. They also are released from collateral (branching) axons of those neurons or through distinct neurons to other parts of the brain, where they act as neurotransmitters; through this separate pathway, they are now thought to influence affect and behavior. Cells respond to the presence of neurohormones when they possess receptors for that hormone; for example, as Dale found, in female mammals the binding of oxytocin to receptors in the uterus during the final phase of pregnancy causes its smooth muscles to contract, facilitating parturition. (Synthetic oxytocin is commonly administered to induce labor in medicalized childbirth.) Oxytocin also facilitates the "let-down" reflex of the mammary glands in lactation. For many decades, these were its only known effects.

The scientific treatment of oxytocin is no longer limited to the mechanics of female reproduction. Oxytocin is now thought to be active in both sexes, implicated in a range of dynamic bodily systems and "widely distributed through the body and brain" (Feldman 2012, 380). It remains heavily identified with the maternal body (whereas vasopressin is treated as a male equivalent) (Willey and Giordorno 2011). However, its agentic capacities have expanded to include a host of potential social roles. Its sociality was first explored in animal research on memory in the 1960s; in this literature, it is roughly associated with increased recall of positively experienced social stimuli and poorer recall of negative social stimuli (MacDonald and Mac-Donald 2010). In the late 1980s oxytocin's roles extended to include sexual behavior when researchers observed an increase in sexual activity of female mice they had injected with oxytocin (den Hertog et al. 2001; Herbert 1994). Researchers now believe oxytocin is released during orgasm as well as during intimate touch and contact.[2] In rodents the release of oxytocin during mating is thought to increase social recognition of partners through its tie to the olfactory system (Borrow and Cameron 2012). As I discuss later, one influential hypothesis is that oxytocin enables the establishment of partner preference in monogamous species of mammals; another is that oxytocin facilitates not only physiological but also emotional processes related to reproduction in pregnant and postpartum mammals. Oxytocin is believed to interact with systems that release dopamine (a neurotransmitter linked to pleasure and reward) and adrenal corticotrophic hormone (involved in the production of steroid hormones in response to stress) as well as memory, and it is associated with learned motivations for attachment and safety. Empirical and theoretical treatments of oxytocin in social neuroscience link it to a wide variety of affective experiences, including trust, suppression of fear, romantic love, and parental bonds (Heinrichs et al. 2009).

A new methodology using the nasal administration of oxytocin has opened up experimental research on oxytocin's effects in humans. Much of this work focuses on trust and cooperative behavior. In a typical study, a group of human subjects is given oxytocin and asked to participate in an experiment designed to measure a positive social affect; these subjects are compared with a control group that completes the task without being given oxytocin.[3] A review by Graustella and MacLeod (2012) summarizes the findings of nineteen such studies: The administration of oxytocin is positively correlated with cooperative and trusting behaviors, facial emotion recognition, and memory for social information; it is negatively correlated with anxiety and blood pressure in stressful situations. Administered oxytocin is thought to improve social cognition, reduce anxiety, and affect motivational states related to affiliation, and its deficiency is being proposed as an explanation for certain social and cognitive disorders, including autism. The effects of administered oxytocin are not predictable, however, and in some experiments they are explicitly manipulated by context. For example, Bartz et al. (2011) cites three studies showing that the positive effects of nasally administered oxytocin "disappear if the potentially trusted other is portrayed as untrustworthy, is unknown, or is a member of a social out-group" (305). Andrew Kemp and Adam Gaustella (2011) point out that in human studies, oxytocin has been linked to not only positive emotions but also envy and gloating.

Like mirror neurons, oxytocin is sometimes treated as if it operates singularly, rather than in intra-action with other bodily systems, and deterministically, as if its properties can explain a particular affect or behavior. For example, one can now find many references to oxytocin as a "love" or "trust" hormone, or even a "moral molecule." This atomistic treatment of oxytocin not only reduces complex, multifaceted, and culturally rich concepts to neurobiology but also obscures the many other neural systems that are thought to be involved in trust, cooperation, attachment, and sexual response, and that are understood to intra-act with the production and utilization of oxytocin, as well as the dynamic relations these systems have with their environments (Zak 2012). Thus, while some researchers and interlocutors of neuroscience attribute affects and behaviors to oxytocin, others argue that its activity in the body cannot be understood independent of context. Oxytocin functions "are extremely diverse" (Van Anders et al. 2013, 1116); oxytocin "may not have a 'function,' and may exert different and even opposing influences on behavior, based upon the sophisticated pattern of neuromodulation in the brain and a particular social arrangement" (Churchland and Winkielman 2012, 398). In research examining the effects of experience on oxytocin production and receptor density (e.g., Feldman 2012; Gil et al. 2013; Phelps et al. 2010; Veenema 2012), oxytocin is treated more as an actant that participates in intercorporeal events and responds to context and experience than as an atomistic or determining entity.

Pair Bonds and Monogamous Voles

The most influential research on oxytocin's role in attachment has been conducted on pair-bonding voles. Voles are a type of rodent native to North America that look like chubby mice. They are considered ideal for studying the neurobiology of social behavior because different subspecies of voles have extremely different social propensities. Like most birds, but unlike the vast majority of mammals, prairie voles are considered to be monogamous. They are described as devoted partners, bonding after a single mating and sharing the parenting of offspring. Animal behaviorists believe that 90 percent of adult prairie voles form long-term pair bonds with a single mating partner (researchers rarely address the 10 percent who do not pair bond). Eighty percent of these reportedly do not pair again after loss of a mate. Prairie voles also coparent, and do so for comparatively long periods. In the lab, pairs strongly affiliate with each other, and when separated, they exhibit symptoms of depression (Bosch et al. 2005; Ophir et al. 2008). Because of their apparent devotion, prairie voles are understood in animal research as models of monogamy. By contrast, montane voles are observed to have indiscriminate sex and no long-term bonds, and they are not biparental. Montane males reportedly do not tend to pups, females affiliate with them only immediately after parturition, and both sexes are generally less social.

In the early 1990s Tom Insel and Lawrence Shapiro (1992) proposed a neurobiological explanation for this difference. They hypothesized that the behavioral disparity was due to differences in the oxytocin system. By looking at postmortem brain slices stained with radioactively labeled oxytocin, they found differences between the two types of voles in the density of oxytocin receptors in two areas of the basal ganglia. These areas are thought to participate in what are considered the reward (dopaminergic) and fight-or-flight (stress hormone) systems of the brain. Insel and Shapiro found that the montane voles had far fewer oxytocin receptors than the prairie voles. They found similar differences in vasopressin.[4] In further studies Insel and Shapiro modified the availability of oxytocin in prairie voles and observed changes in behavior. They reported that prairie voles whose oxytocin receptors were chemically blocked behaved more like montane voles, mating but not establishing partner preferences, whereas prairie voles injected with oxytocin showed long-term bonding even when they were prevented from mating. They concluded that oxytocin "appears to be both necessary and sufficient for partner preference formation, the first step in the development of a pair bond in this monogamous species" (Insel et al. 1997, 34).

What relevance do studies of voles have for understanding human kinships? Like 95 percent of mammals, humans are not considered a monogamous species. An oft-cited reference is Murdoch's *Atlas of World Cultures*, in which only 17 percent of the 563 cultures are listed as having monogamous kinship structures. Despite this, prairie voles are often treated as a model for human romantic love (Ophir et al. 2008, 1144). Some researchers argue that even in nonmonogamous societies, humans are capable of sociosexual bonding akin to those of prairie voles, and that such bonds are "important to the physical and mental health of individuals and their children" (Young et al. 2011, 54). In a review in *Frontiers of Neuroendocrinology*, Kimberly Young et al. explain: "Intense attraction between mates, often referred to as romantic or passionate love, is one of the most powerful forces driving human social behavior, and often precedes the formation of enduring, selective attachments between sexual partners (i.e., pair bonds). Although such sociosexual attachments are most prevalent in industrialized cultures with a monogamous social organization, they occur in nearly all human societies, regardless of subsistence mode (e.g., pastoralist, agriculturalist, etc.) or mating strategy (e.g., polygamy and monogamy), and are therefore an intrinsic part of human social behavior" (2011, 54). The authors argue for the functional significance of pair bonding in humans, pointing to cross-cultural measures of health and longevity of monogamously paired individuals, as well as offspring of such pairs. In accordance with much of the literature, they assume pair bonds are heterosexual and reproductive, and that such bonds confer evolutionary advantages. These assumptions, which depend on a "specific cultural position" that monogamy is "universally desirable" (Van Anders et al. 2013, 1115), exclude the range of felt bonds that might otherwise fall under the description of sociosexual attachments. This exclusion does not merely limit the relevance of the research but also constrains the thinking about the neurobiology of attachment even within reproductive relations, a position I elaborate later.

Maternalizing the Brain

The social neuroscientific explanation of oxytocin as an actant in human kinships begins in evolutionary terms not with pair bonds but with mother-infant bonds. Oxytocin is strongly identified with maternal bonding with offspring. For example, nonhuman animal experiments have explored the effects of both increasing and blocking oxytocin on females' interactions with their offspring. Researchers have chemically blocked oxytocin receptors, and genetically modified oxytocin genes, of mice and rat mothers, and reported the diminishment of expected behaviors toward their young, such as licking and grooming, nest building, and prolonged feeding (Ferguson et al. 2000). Conversely, animals whose oxytocin system is upregulated have been observed to increase such behaviors, and also offer nurturing to young that aren't biologically related.[5] Oxytocin is also thought to have a role in alloparenting (care of nonrelated young) and male attention to offspring, but research on oxytocin's role in parenting has overwhelmingly focused on mother-infant bonds.

Some argue that the participation of oxytocin in maternal bonds is an effect of evolutionary adaptation. The gene for oxytocin initially ensured only that females, as Patricia Churchland argues, "had the resources and motivation to suckle, defend, and more generally, to devote herself to the welfare of her helpless juveniles until they were independent" (2011, 31). Through genetic changes, the social role of oxytocin expanded beyond mother-infant bonds. The modification that "yields caring for others that are offspring could be further modified, perhaps in quite minor ways, to yield caring for others that are not offspring, but whose well-being is consequential for the well-being of oneself and one's offspring" (32). Across species of mammals, varieties of social arrangements may reflect different underlying patterns of oxytocin receptors and related circuitry. A strong view claims that, in humans, these biologically supported attachments are the scaffolding for all social bonds and for social values (e.g., Young and Alexander 2012).[6]

Many theories of attachment see mother-infant bonds as a touchstone for all forms of bonding, where the "behavioral building blocks of maternal affection—gaze, touch, voice, and affect—serve as the basic channels for the expression of love that underpin any form of human intimacy" (Feldman 2012, 383). In one hypothesis, articulated by Churchland (2011) and Larry Young and Brian Alexander (2012), mothers are neurally wired for affectionate behaviors. Specifically, a *maternalized brain* underpins maternal affection. This argument builds on research by Jay Rosenblatt, considered the "father of the experimental study of maternal behavior," on the relation between the maternal hormonal cycle and behavior (Fleming 2007, 8). Maternalization roughly describes a set of processes that are believed to occur mainly during pregnancy, parturition, and lactation. During these events, surges of estrogen, progestin, prolactin, and oxytocin prepare the body for physical labor. They also prepare the brain and mind for affective labor. Churchland and Young both argue that oxytocin's primary social function is to orient a female toward her offspring so that she is willing to nurse them, protect them, and generally provide for their well-being. As Young explains in a *Time Magazine* interview, "the [oxytocin] molecule was first involved in the physiological process of birth. With mammals, you've got them needing to nurse and oxytocin is involved in milk ejection. You also have to transform the mother's brain so that she focuses her attention on the baby, so she becomes a mother. Oxytocin is acting both on the body and the brain to transform her into a mother" (Szalavitz 2012).

According to the "maternalized brain" theory, this transformation begins even before birth, independent of any social interaction with offspring. It is a multistage process. First, through surges of prolactin and estrogen, pregnant females are oriented toward a state of maternal expectation; they become absorbed in preparations for birth and caretaking. Churchland explains, "In rodents and cats, for example, this causes the pregnant female to eat more, to prepare a nest for the expected litter, and to find a place reckoned as safe to give birth" (2011, 33). In humans, Churchland claims, "females too respond to a 'nesting' urge as the time for delivery draws near, and (as I can personally attest) begin energetically to house-clean and finalize preparations for the new baby . . ." (33). Next, parturition involves a significant increase in the density of oxytocin receptors. The peripheral circulation of oxytocin allows the uterus to contract; the parallel release of oxytocin in the brain "triggers full maternal behavior, including preoccupation with the infants, suckling, and keeping the infants warm, clean, and safe" (33). Finally, postnatal interaction builds on this foundation. Like parturition, lactation has not only peripheral but also central nervous system effects. The stimulation of nipples during nursing promotes not only milk let-down in the mammary glands but also reduced activation in the stress system of the brain (Young and Alexander 2012, 98). Lactation has "pleasurable and calming" effects due to the release of oxytocin and dopamine (Churchland 2011, 34).

Other kinds of maternal-offspring interaction also involve feelings of pleasure and reward. With a hypothalamus primed by prolactin and estrogen, interaction with offspring stimulates the release of oxytocin from the pituitary gland, which in turn stimulates the release of pleasure-inducing dopamine via the nucleus accumbens. This provides a feeling of reward. As Churchland puts it, "When a mammalian mother is successful in making the infant safe and content, endogenous opiates as well as oxytocin are released, both in the brain of the contented infant, and in the brain of the relieved mother. Being together feels good" (40). This cycle of social interaction and dopaminergic reward establishes what Young calls an "appetite" for contact. In rats, for example, the smell and sound of young are "now so attractive that a new rat mother will cross an electrified grid to fetch one" (Szalavitz 2012, n.p.). In sheep and humans, too, the dopaminergic reward encourages caretaking. "The same dopamine-reward pathway is engaged when a mother senses the sight, smell, and sound of her own baby, linking the sensory cues, the emotional feeling, and the reward, and muting the prefrontal cortex, all of which motivates mothers to nurture. Caring for a baby feels good, especially if it's your baby" (Young and Alexander 2012, 102).

What of kin attachments that fall outside the birth mother–infant dyad? Churchland believes that because of the plasticity of the oxytocin system, the attachment for adoptive mothers "can be every bit as powerful"—in other words, as biologically supported—"as attachment to a baby carried and delivered" (2011, 34). Young's explanation is that while pregnancy, birth, and lactation are not absolutely necessary for bonding, a female brain is. His account is informed by brain organization theory, which argues that brains are sexually dimorphic. He holds the view that all female mammals have existing circuitry for mothering that is activated by pregnancy (Young and Alexander 2012, 95). As an example, he cites experiments by Rosenblatt on maternal behavior in rats. In the first study Rosenblatt (1967) placed reproductively naïve female rats together with neonate pups. The adult females' first reaction to them was either avoidance or aggression. In a few days, however, they began to approach the pups. After about a week, they developed species-typical maternal behaviors toward the pups, crouching above them as if to nurse, licking and grooming, and retrieving them when they were separated. In a follow-up study Terkel and Rosenblatt (1968) infused the bloodstreams of virgin female rats with blood from rats in late-stage pregnancy. Then they repeated the experiment as before. These blood-doped rats immediately engaged in maternal behaviors. The researchers attributed this difference to the infusion of pregnancy-related hormones. Young interprets this as evidence that "the female rat's brain contained the wiring necessary for her to behave like a mother without actually being one" (Young and Alexander 2012, 95).[7]

There are several problems with this argument. Among them, the sex dimorphism Young describes is disputed by Rosenblatt's study of male rats, which finds that, under certain conditions, "the basic capacity for maternal behavior is present in both sexes" (Wong 2000, 71). This raises the question of how behaviors are understood as maternal or paternal in the first place. Angela Willey and Sara Giordorno (2011) argue that Young's lab has focused overwhelmingly on oxytocin and female maternal behaviors, while paying relatively little attention to oxytocin's role in male parenting; instead, the researchers attribute to vasopressin what they see as male behaviors, such as mate guarding and the protection of pups. Yet all male prairie voles produce oxytocin and also nurture their young, and females exhibit some of the same behaviors that in males are attributed to vasopressin, such as protectiveness and aggression. Willey and Giordorno argue that a framework of sex/gender difference constrains the interpretation of animal behavior in these accounts. Another problem is that Churchland and Young also offer highly idealized impressions of the affective, embodied maternal experience, as if, outside of pathological cases, all pregnant females are consumed with thoughts of or preparations for future offspring, lactation is generally calming and pleasurable, nurturing is automatically offered and consistently feels good, and interest of mothers in offspring is always intense. But this depiction does not hold up to much scrutiny.

To take just the case of lactation: While there is some evidence that lactating mothers experience a reduction in stress and an increase in dopamine, this fact alone does not even begin to describe the experience, for humans at least. As even the most cursory look at breastfeeding research suggests, the experience is widely variable and involves many social, cultural, economic, physical, and psychological factors (MacLean 1988; Thulier and Mercer 2009). Historian Londa Schiebinger (1993) notes that since antiquity, many women have practiced breastfeeding as a form of domestic labor, whereas other women have considered breastfeeding their own children a great burden.[8] Colonials utilized the services of native women and slaves, urbanities engaged wet nurses from the country, and upper- and middle-class families employed the poor. Currently, women's reticence to breastfeed is viewed as a global health crisis. Even in countries where the majority of birth mothers now initiate breastfeeding, such as the United States and United Kingdom, most do so for much shorter than the recommended periods, and many abandon the practice earlier than initially planned (Andrew and Harvery 2011).[9] There is a considerable public health literature addressing the reasons for women's reluctance to breastfeed. In addition to economic and social factors, women describe a multiplicity of psychological and physical reasons. While some mothers, like Churchland, have positive experiences with lactation, others describe the practice as trying, difficult, painful, disruptive, unpleasant, or even violent (Hurley et al. 2008; Kelleher 2006; Schmied and Barclay 1999; Thomson and Dykes 2011; Tucker et al. 2011). Some also note the incongruity between naturalizing discourses about breastfeeding and their real-life experiences, which can foster feelings of failure or guilt (Burns et al. 2010; Mozingo et al. 2000; Schilling et al. 2008). Among those who report success, breastfeeding is not generally described as seamlessly, effortlessly, or automatically harmonious. Instead, mothers portray it as a learned competency (Schilling et al. 2008), one that involves a "complex interactive process" (Leff et al. 1994, 99) of gradually developing a successful routine through trial and error.

Biobehavioral Synchrony

The adaptionist theory argues that a maternal-infant bond has both its quality and its object determined by evolution. Alternatively, embodied relations can be understood as processes that develop in specific temporal, spatial, and interactive contexts, with varied outcomes. Ruth Feldman takes this view to argue that hormonally supported affective bonds develop through close interpersonal interaction. She conceptualizes this in terms of *biobehavioral synchrony*. *Synchrony* refers to the idea that bodies in close relationships become attuned to each other's sensory, motor, and behavioral cues. Synchrony highlights "the time-based component in interpersonal encounters and emphasizes the ongoing organization of social behavior into repetitive-rhythmic sequences" (Feldman 2007, 340). Intraspecific synchrony has been observed in animals, for example, in group actions of birds and fish that demand micro-coordination. Humans, too, are believed to exhibit synchrony, for example, when they unknowingly coordinate eye blinking and speech patterns (Argent 2012). As Hall put it, "People in interactions move in a 'kind of dance,' but they are not aware of their synchronous movement" (1976, 71, cited in Argent 2012, 116). Synchrony has been used in developmental psychology to address the mutual adjustments of bodies in parent-infant interactions. Multispecies scholars are now examining synchrony between animals of different species, for example, between humans and horses and humans and dogs as they move together. Gala Argent, for example, argues that the coordination of horse and rider involves a "corporeal, synchrony-induced sensation of boundary loss," which can be experienced as pleasurable for both (121).

In specifically *bio*behavioral synchrony, Feldman looks at the physiological components of close relations and their role in building attachments. Ongoing, repeated practices of intimacy and care, such as those that take place in breastfeeding, involve "micro-level social behaviors in the gaze, vocal, affective, and touch modalities" (Feldman 2012, 380). Feldman argues that these are "dynamically integrated with online physiological processes and hormonal response to create dyad-specific affiliations" (380). In other words, she argues that while attachments are supported by oxytocin (as well as other systems), they are not automatic or static. Rather, bodies in interaction build them. During touching and contact, for example, the oxytocin system offers biobehavioral feedback in two directions; parental touch, for example, triggers oxytocin release in both infant and parent. Over time, "these discrete synchronized bio-behavioral events coordinate to form the unique bond that characterize the rhythms, content, focus, and pace of the specific attachment relationship . . . . Synchrony, therefore, describes a critical component of close relationships that builds on familiarity with the partner's style, manner, non-verbal patterns, personal rhythms, behavioral preferences, and pace of intimacy" (382). Here, synchrony is an affective, material event in which bodies become mutually attuned and oriented toward each other. It is simultaneously experienced in individual bodies and wholly relational, generated through embodied social interaction with specific others. In Feldman's view there are long-term effects; she argues that an individual's capacity for affiliative bonds throughout life is built on "unique neurohormonal systems and brain circuits" that are tied to specific behaviors taking place in social interactions (381). While Feldman privileges the nursing dyad, synchrony and its neurohormonal aspect might be explored in any close relation. Feldman's lab, for example, has examined neurohormonal involvement in best friendships as well as between fathers and infants and partners who coparent.[10]

As felt affiliations that are built through intercorporeal experience, Feldman argues that attachments are better understood not as the effect of essential traits, but rather as ongoing practices of bodies that occur in specific temporal, relational, and environmental contexts. While some researchers are committed to biologically determinist theories that emphasize constraint, the focus on affective kinship also enables the exploration of bonds through the lens of bodily capacity. It opens up the biology of kinship to social interaction, intercorporeality, and intersubjectivity. It is not too far from here, I venture, to conceptualizing attachment as materially as well as discursively performative. As *materially* and not only symbolically performative, attachment bonds can be seen as embodied structures of experience that, even while they continue to change, have effects that condition future experience. This view demands both an embodied and, as I argue, a potentially queer way to think about what biological relatedness is, and of the implications of its loss, disruption, or erasure.

Rethinking the "Complex" Bond

Many scholars in the humanities and social sciences have argued that opening up the brain to social interaction, affect, and intercorporeality enables us to move past biological determinism in thinking about the nature/culture relation. To these arguments, I want to add that it also could help us move past the relentless heteronormativity of biological stories about kinship. In that vein, I want to return briefly to the treatment of prairie voles, which are viewed as nature's standard for monogamy among mammals and have a significant role in neurobiological knowledge of kinship. Much of this research is guided by the "evolutionary assumption that each individual organism (human or otherwise) has as its primary 'goal' the perpetuation of its own genetic material through reproduction" (Willey and Giordano 2011, 112). This means that all bonds—sexual, companionate, filial, as well as maternal and paternal—are firmly in the service of reproduction. Further, it means that in species that form pair bonds, sexual, maternal, and paternal attachments are inextricably tied together. The formula works like this: Mating generates orgasm, which releases oxytocin; oxytocin initiates a cascade of effects in the brain; the result is pair bonding and reproduction. Reproduction, as we saw earlier, also involves oxytocin output. At each stage, neurohormonal and other systems work to bind these behaviors together. Some researchers have argued that among prairie voles, a single mating can lead immediately to pair bonding, reproduction, and lifelong monogamy. This packaging of bonds reportedly confers evolutionary advantages. As Kimberly Young et al. argue, "The co-occurrence of these behaviors in pair-bonded individuals makes sense when viewed through the lens of evolutionary theory, which suggests, in part, that pair bonding became adaptive under conditions in which additional parental investment was required to ensure the successful rearing of young. Indeed, the same selection pressures that necessitated the presence of both parents for offspring survival would likely facilitate the formation of a partnership between mates and mechanisms through which to maintain this partnership (e.g., mate-guarding)" (2011, 54). For prairie voles, a "complex social bond," as Young et al. (2011) put it, means something like a nuclear family. Tied to assumptions of the inherent heterosexuality of individuals and the functionality of monogamy, this perspective leads to a deterministic vision of their kinships.

It might be best in the face of this research to declare the incommensurability of nonhuman animal bonds and human kinships. Instead, however, I want to highlight scientific practices that call into question the heteronormative depiction of nonhuman kinships. These begin with disentangling the associations between mating, bonding, reproduction, and monogamy. In mammalian research *mating* is generally understood as male-female reproductive coupling, and *pair bonds* are commonly defined as partner preferences that include heterosexual mating, behaviors like licking and grooming, and cohabitation. They also include "the subsequent development of selective aggression toward unfamiliar conspecifics, and the bi-parental care of young" (Young et al. 2011, 53). *Monogamy* is generally defined as the formation of exclusive pair bonds. This commonly means that only male-female reproductive behaviors are counted as matings, only opposite-sex couplings are counted as pair bonds, pair bonds are assumed to be the result of heterosexual mating practices rather than cohabitation, and, for species like prairie voles, cohabitating pair bonds are assumed to be sexually exclusive.

However, a sea change is taking place in animal behavioral research on the subject of monogamy. In the past two decades new methodologies including genetic testing have transformed many assumptions about monogamy and other sociosexual behaviors in animals. As Ulrich Reichard summarizes, recent findings on monogamy in birds and mammals suggest that "social, sexual, and reproductive relationships are complex and varied, even in socially monogamous species where hitherto 'monogamous' appeared the only term necessary to describe structures resembling the nuclear family" (2012, 66). The coupling of sex and pair bonding, pair bonding and fidelity, and reproduction and caretaking are being challenged. In avian research the consensus on Lack's famous declaration in 1968 that 90 percent of birds are monogamous has been overturned (Neudorf 2004). The use of genetic testing in the 1990s led to the surprising finding that many birds co-nesting for life often were rearing a group of offspring with diverse parentages. Reichard notes that less than 25 percent of "monogamous" birds studied practiced *genetic* monogamy, as opposed to *social* monogamy. The former refers to sexual exclusivity and the latter to exclusive pair bonding. Genetic monogamy is now considered to be more rare than once suspected in mammals as well. "Behavioural observations [now] show that living together in social monogamy does not equate to monogamous mating or reproduction. Thus the females of the Lesser apes, Alpine marmots, fat-tailed dwarf lemurs, aardwolves, common marmosets and small Mongolian gerbils, and many pair-living birds are not too particular about sexual fidelity to their male social partners" (2012, 63). When partner preference and sexual practices are disentangled, animal social patterns become considerably less predictable.

Surprisingly, not until 2008 was genetic monogamy tested in prairie voles, but the results were similarly disruptive. At a field lab in Tennessee, biologist Alexander Ophir and his colleagues put radio collars on ninety-six prairie voles and tracked their movements in their native habitat. After two weeks, they identified pair couplings by finding patterns of cohabitation. They also genetically tested pregnant females. The researchers report that the voles did make pair bonds, but the couples did not exhibit sexual fidelity. Instead, many of the voles mated with multiple partners while cohabiting with one. Ophir et al. (2008) argue that both females and males have sex with voles to whom they are not pair bonded, and males rear offspring to whom they are not genetically related. Reporting on this study, the news section of the journal *Nature* ran a headline worthy of any tabloid: "'Monogamous' Voles in Love-Rat Shock." The account reads: "By traditionalist standards, prairie-vole couples may enjoy the ideal relationship: the rodents form lifelong partnerships—a highly unusual practice in mammals. Males help raise the children; females help build the nest. As for their sex life, let's just say it far exceeds the efforts required for procreation. But the respectable public behaviour of North American prairie voles (*Microtus ochrogaster*) may hide a bed-hopping double life. Paternity tests published last week indicate that the animals touted as paragons of monogamy frequently cheat on their partners" (Ledford 2008, 617). This finding not only damages the reputation of prairie voles as pillars of monogamy but also complicates evolutionary theories that use them as a model to seamlessly link sex, reproduction, and kin bonds.

There is also the related problem of the heteronormative gaze in animal research. While same-sex behavior has been observed in many species and well documented in about five hundred species (Bagemihl 1999), researchers have not seriously considered the implications of this for theorizing kinship until recently (Bailey and Zuk 2009). Thus, various kinds of pairings, even those that take place in the lab, can be misrecognized or ignored.[11] Bailey and Zuk note a general tendency in animal research to reassign "behaviors that would normally be expressed in the context of an opposite-sex courtship or reproductive interaction, but are instead co-opted for another function" (2009, 444). They argue that the prevalence and diversity of same-sex interactions in animals calls for rethinking evolutionary theory. Rather than explaining these separately from other sexual behaviors, they call instead for treating them as "potential selective agents in and of themselves" (445). The sex practices, pair bonds, and paternal styles of prairie voles and other nonhuman animals seem to have multiple and even cross-purposes, rendering any singular, unifying evolutionary explanation for all affective ties improbable.

Queer Ties and the Embodiment of Kinship

Social neuroscience is making a strong case for the relevance of intercorporeality and affect in attachment and kinship. Much of this work, unfortunately, adheres to the "traditionally constructed linear equation of sexual reproduction" (Hird 2004, 222), where heterosexual coitus leads almost inevitably to pair attachment, pregnancy, and maternal bonding. To the extent that social neuroscientists rely on this model, they cannot even begin to address everyone's experiences of attachment and kinship. The model poorly explains the appearance of other kinds of attachments in scientific research—paternal, alloparental, and same-sex bonds, to begin with—and fails to consider how nonreproductive, nonheterosexual, or otherwise queer kinships in human lives are biologically enabled. Even kinship structures that do fit more or less neatly into a reproductive equation—such as mother-infant nursing dyads—cannot be adequately described in universal and essentialist terms. Affective ties are experiential and, in human contexts at least, situated in cultural and historical circumstances, including relations of exploitation and domination. A reductionist approach leads to erasure of this specificity, to the bracketing of context, and to ignorance of the effects of discursive, economic, and political structures on embodied relations. If social neuroscience is getting kinship wrong, then, it is not because it looks to the biological body. Rather, it is (in part) that its understanding of the body is constrained by heteronormative logic, a logic that does not merely exclude some bodies in favor of others, but obscures the complexity of the experience it is trying to explain for *all* bodies.

The erasure of queer experience, however, has particularly violent effects for people whose bonds are deemed to disappoint nature. Butler (2004) argues that kinships not only depend on, but are also constituted by, social recognition, which is conferred only on certain classes of relations, most ideally, heterosexual, two-parent families. As Butler notes, being recognized as legitimate kin means that "everyone must let you into the door of the hospital; everyone must honor your claim to grief; everyone will assume your natural rights to a child; everyone will regard your relationship as elevated into eternity" (111). The lack of social recognition of queer kinship bonds threatens them, as when "the sense of delegitimation can make it harder to sustain a bond, a bond that is not real anyway, a bond that does not 'exist,' that never had a chance to exist, that was never meant to exist" (114). Because, for Butler, bonds are actualized through social recognition, the very reality of unintelligible bonds is put into question by heteronormativity. But as Butler also points out, the social recognition of some queer bonds threatens others, those that veer farthest from reproductive relations. At best, the legitimation of some bonds forecloses other possibilities; at worst, it can lead to "new and invidious forms of social hierarchy" in which some bonds are vilified in the effort to give social status to others (115). It is these prospects that cause Butler to weigh the limits to kinship altogether. (Perhaps, she suggests, kinship should be abandoned in favor of bodies and pleasures.)

Neuroscientific moves to deemphasize genetics in favor of affective ties take place in the context of a postgenetic world where kinships can no longer be captured by genetic relatedness. Even as the biological sciences acknowledge, and facilitate, the reality of new reproductive and nonreproductive modes of life and relation, social hierarchies are being reinscribed. Neurohormonal accounts of kinship may make possible new hierarchies that center on the quality of bonds purportedly afforded by neurohormones. For example, the status of mothers who undergo IVF with egg donation, and give birth to babies not genetically their own, could trump those who use surrogates to birth their genetically related offspring. Adoptive parents of infants, who have more time to build bonds early in development, may find more validation than those who adopt older children. Mothers who lactate may be given more approval than parents who do not or cannot. (Social neuroscientific research is now attempting to address the relative strength of bonds between mother-infant dyads with different histories of birth and feeding.) Conceivably, those with more oxytocin receptors could be measured as having more resources for attachment than those with fewer. Or, oxytocin will become available as a method of facilitating such bonds. Social neuroscientists are already proposing theories about oxytocin's role in mental, cognitive, and social disorders, including autism, and advancing administered oxytocin as a drug therapy. The scientific measure of certain bonds as neurobiologically supported, undoubtedly, opens up intercorporeality and affective kinship to biopolitical intervention and regulation.

To contest neurohormonal reductionism, in my view, it is better to insist on the queer potential of nature, and nature/culture, than to deny the imbrication of the biological body in lived experience. Social constructionist accounts of kinship, which depend on language and human exceptionalism, underestimate what relational ontologies are actually at stake when they are threatened by erasure and misrecognition. On what basis should one demand acknowledgment at the hospital? What loss should be recognized at the funeral? The kinship bonds that so many people want acknowledged—those that are dismissed and sometimes obliterated by the heteronormative gaze—are not solely representational. They are not exclusively discursive, nor even exclusively human. The kinships worth defending involve, among other things, the lived, felt orientations of bodies toward each other (Eng 2010). When kinships are lost or denied, it is in part bodies that are at stake, bodies that have been transformed in various ways by unique, intercorporeal relationships of intimacy and care, and thus made vulnerable to their denial or loss (Butler 2010). What is needed, ultimately, is a rethinking of what it means for humans as well as nonhuman animals to be related, including what it means to be biologically related. Being biologically related does not have to mean genetically related; it can mean *having a biological investment in another*, in the form of an intercorporeal tie to another, that is the product of interaction, intimacy, or companionship. The transformation of bodies as they live with and toward others is a kind of relatedness that ought to be recognized. For humans, at least, such relatedness is also discursive, historical, and political. An intercorporeal perspective may help us deepen our sense of what it means to say that, as Eng (2010) argues, affective kinships are woven through with gendered, racial, national, and transnational histories, because it indicates that these histories in some way are materialized in embodied experience.

Conclusion

The Multiplicity of Embodiment

Biological stories about human (and nonhuman animal) experience do harm when they deny the complexity, specificity, and multiplicity of lives. For example, as the foregoing discussion shows, the imposition of hetero-normativity not only essentializes brains and body-subjects according to reproductive norms but also belies social bonds and events that lie outside its gaze. The exclusions it achieves are many, from same-sex couplings that are misrecognized or deemed to be less than natural, to parenting practices that defy adaptationist logic, to intra- and cross-species attachments that depend on bodies but cannot be explained in sexed/gendered terms. My argument for recognizing this heterogeneity builds, of course, from feminist critiques of the sciences of sex/gender. Feminist scholars and scientists have been contesting claims that brains are prenatally sexed—claims that, in their expansive versions at least, seem to be undeniably ideological—for at least three decades. The sexed brain research, while ostensibly recognizing difference, in fact tends to generate uniform, generalized accounts of male and female brains, traits, and behaviors, thereby obliterating the variations within each category, as well as those variations that call the two bifurcated categories into question. Such accounts are unfathomably resilient, in spite of their incompatibility with the plasticity and dynamism now being attributed to neural matter. Critics of sexed brain research have argued that scientific representations produce the objects they purport to merely observe. Beyond this, some have challenged the power of science to declare the truth of the subject as emanating from biology, whereas others have championed alternative models and methods that recognize biological variation, dynamism, and multiplicity.

The latter strategy is important because it has become clear that revealing the bias of scientific representations is not, in the end, enough. If critique is limited to addressing the historicity of scientific claims or unveiling their normative character, the biological materialities themselves are left by the wayside, to be picked up by those who do not feel queasy about finding truths in nature. The problem is not merely that truth claims are being made—that, in their hubris, scientists forget Thomas Kuhn's lessons in *The Structure of Scientific Revolutions* (1962) and treat their findings as unadulterated facts. The issue is also that brains and bodies and lives will be changed, partly in relation to how brain knowledge is enacted. For example, neuroscientific practices that address attachment will affect the clinical management of reproduction and could also produce oxytocin as a prosocial drug. Particular enactments of biosocial plasticity may inform educational policies that literally target brain regions, and populations, for modification. It would do little good to dismiss hormones, or brain plasticity, as immaterial constructs, or as representations that matter *only* because they are symbolically meaningful. An adequate response to these phenomena must not only recognize the power of scientific knowledges to create objects or to produce normative models for body-subjects—for example, good students and citizens, ideal mothers, or reproductive families. It must also acknowledge the material specificity of bodies, their ability to change and be changed, which is entangled with (but not reducible to) how they are understood. The stakes, then, are onto-epistemological. They involve how neurobiological bodies are known as well as what they are, what they can be, and what they can do.

The inadequacy of exclusively representational thinking is revealed when one is forced to attend seriously to a particular body. One can say (to use a stark example) that neurological accounts of Phineas Gage's brain injuries underwent revisions as the theory of localization waxed and waned. With the benefit of historical perspective, it is easy for scientists and clinicians to admit, "our ideas about the function of the frontal lobe have had a torturous and ambiguous history and are still far from clear" (Sacks 1995). Indisputably, though, Gage had an iron rod thrust through his skull. So, while it is necessary to address the varied ways neurologists understood his brain injury, it is not reasonable to say that they had no business thinking about its effects on his brain, and even on his personality and behavior. Similarly, attending adequately to illness often involves grappling with bodies as unavoidably material realities. Bruno Latour, the sociologist most known for showing how scientific objects are constructed, had colon cancer, as he mentioned in his essay on the limits of social constructionism (2004). This he'd rather not regard merely as a "fact" to be deconstructed. Catherine Malabou's grandmother had Alzheimer's disease, which brought her to thinking about destructive or "negative" plasticity as inescapably material (2012). Others writing about bodily matters, including illness, disability, and even experiences of oppression and marginalization, have also challenged exclusively representational and (de)constructionist approaches. If scientific facts and medical models do not fully grasp such embodied realities—if even they do violence to them—epistemological critique only gets one so far as well.

This argument, however, should not be taken as a concession to some ultimate or inevitable biological reductionism, as if the social addresses only the surface; to really get at the depth one has to give in, finally, to biological nature. The point, rather, is that these two shouldn't be easily separated. The example of illness and injury can be disarming, leading one to put down the tools of critical thinking altogether. *Of course,* Latour's readers might well reason, *he must treat his cancer as biologically real, and not merely as a construct, or a metaphor. What good would that do?* Or: there is no doubt, Malabou intimates, that Alzheimer's disease can create unbearable suffering. The mere thought of such suffering might be taken for evidence that biology trumps all. But while cancer and Alzheimer's disease are absolutely biological, they are not strictly, purely natural, if that means untouched by the contingencies of social life. They involve undeniable disruptions, genuine pain, and irreducibly physical events (lesions, malignancies, plaques, and tangles), but their etiologies are thought to involve both biologically and socially patterned exposures and vulnerabilities, and they are experienced not as pure processes of nature, but as entangled with the practices, meanings, and knowledges that variously and differentially shape embodied lives. To take a *stubbornly realist attitude,* as Latour put it, is not to deny these social imbrications, but rather to see them, too, as actual. Similarly, to say that oppressions are felt and experienced in the body—to say that they have fleshly, biological import—does not imply that they are natural as opposed to historical. Nor should it lead to biosocial reductionism and determinism, where social inequalities inscribe the biological body and fix themselves there, crystalizing into forms that determine experience (see Schmitz 2012). A real, actual body, a body that experiences real illness or actual oppression, cannot be treated as purely natural, nor as purely symbolic. Rather, it must be recognized in its material-discursive specificity, or, to put it another way, as complexly embodied (Siebers 2008).

Although bodily problems and vulnerabilities can force one into a realist or materialist mode, the positive capacities of the body also demand attention. After poststructuralism and its privileging of discourse, much contemporary social thought is approaching nature, biology, and matter with a sense of enthusiasm, wonder, or even vitalism. Within this scope of thought, the turn to affect, in particular, is informed in part by depictions of the brain and nervous systems working at levels before and below cognition, which allows some of what has previously been defined as cultural and linguistic to be understood as biological and material. Experiences that are inadequately grasped in language may be attributable to the capacities and excesses of the neurobiological body—its multiple material, social, and intercorporeal agencies. Affect is social, goes the theory, without being derived from the symbolic; bodies can therefore be communicative, or share valences, outside of subjectivity and consciousness. (Thus affect is not exclusively human.) In many accounts of affect the body's vulnerability to power is similarly extended. Affect theorists, along with queer and feminist phenomenologists, underscore how their orientations toward and entanglements with others can change bodies. Thus attention to the body's capacities can help to address in ontological or realist terms the heterogeneity of bodily experiences, even (or especially) those queer affects that have been deemed scientifically, legally, or socially implausible.

If contemplating illness and pain can lead to critical fatigue, theorizing the biological body's dynamic and social propensities can also foreshorten epistemic cares. In response to criticisms that affect theory draws too uncritically from brain science to ground its antirepresentationalism (Blackman 2012; Hemmings 2005; Leys 2011; Papoulias and Callard 2010), some affect scholars point to the dynamism of neurobiology, much as some use neural plasticity as immanent critique of the sexed brain. The concepts of plasticity, biosociality, and embodiment can, in principle, rectify the reductionism and determinism associated with the neurosciences. My own inquiry was animated by the possibility that attention to neurobiology—as envisioned in an era of plasticity, epigenetics, and biosociality—can open up, rather than close down, the material and historical specificity of bodies and lives. I hoped, in other words, to grasp meanings and neurobiological bodies together in order to make more room for the whole range of actual, existing body-minds and body-subjects—especially those that are historically ignored or pathologized by science—to "have a chance" (Haraway 1988, 580). If this is the promise of the biosocial, plastic brain, it is unevenly realized in the examples I explored here.

The plasticity of the brain presumably releases it from biological determinism and enables experience to make its mark. But the implications of this, I argue, often are contingent upon the particular ways plasticity is understood and enacted. The adolescent brain, for example, is not necessarily less determined than it was before its renewed ability to sprout and prune was discovered. Plasticity is sometimes treated as more or less intrinsic in adolescent brain research, such as when the goal is to map differing degrees of maturation across multiple brain regions at given moments in development, rather than to explore the ways experience diversifies brains *and* their developmental trajectories. The result is a universal, mildly pathological adolescent brain, with an immature prefrontal cortex that is no match for its amygdala. This brain is being proposed as a biologically determinist explanation for various social problems involving youth. Those invested in biosocial plasticity might be tempted to say that this research is simply not plastic *enough*, or that it explores *the wrong kind* of plasticity. The research program I discussed on the neurobiological impact of poverty, however, explores extrinsic, experience-dependent plasticity, to questionable effect in my view. Here the aim is to understand the biological effects of poverty, as measured by socioeconomic status, on developing brain systems. To recap, the general hypothesis is that the brain shows distinctive and identifiable neural patterns in response to the multiple negative experiences (stress, nutritional deprivation, toxins, drugs, and so on) that tend to mark the lives of children who are poor more than others. In this research program subjects are differentiated by (bifurcated, gross) measures of class status, as a proxy measure of their exposure to a range of physiological and other factors correlated with poverty. Particular neural processes assigned to specific brain regions are elicited, compared, and found to act (in some ways) differently. These neural differences, sometimes irrespective of their relation to cognitive performance, are interpreted as neural inequalities. Here, the plastic, biosocial brain gains a class phenotype through biosocial reductionism.

The concern I have about this research is not that it sees social inequalities as having biological import. Social inequalities and power relations *should* be brought to bear on how we think about the brain, mind, and body, and neuroscientific accounts need to be cognizant of the different ways that power and inequality can affect lives. But *how*? What sorts of measures are appropriate for grasping this in its complexity? When a childhood is marked by economic vulnerability, chronic stress, and food insecurity, the question is not whether this is experienced neurobiologically. The questions are many, including, How generalizable is that experience, and for whom? How generalizable are its effects, and in what contexts? With what mediating factors? Across what temporalities? To what degree are these captured by existing methods? Bifurcated measures of class difference are not only essentializing; they also presume a singular norm against which variances can be measured, while ignoring the relations between the variances. (How is addressing "poverty," for example, different than addressing "inequality"?). Most worryingly, they enable practices of neurogovernance— such as the targeting of the prefrontal cortices of poor and often minority children—that are justified with reference to social problems that are woven through with classism and racism. To be clear: The "poverty brain" is far from a mere social construction. Rather, it is a particular enactment of neural difference that has real, material effects.

When social differences are neurally instantiated, they do not cease to be social and situated. Feminists have argued that the social institution of gender affects brains, largely in response to neuroscientific observations of neural sex differences. However, they generally refuse to transcribe essentialized categories onto neurobiological bodies. Instead, they argue for disentangling aggregated properties and traits, which often are lumped together in bifurcated categories, in order to see their complex, overlapping distributions across multiple categories of research subjects and experimental contexts (Rippon et al. 2014). On this account, gender is not a biosocial essence or neural identity. Rather, it is materially and discursively performative, gaining self-identity only in its contextualized, repeated expressions of difference, which are not independent of, but affected by, their scientific observation.

Plasticity, then, is not precisely equivalent to *becoming*. Similarly, "embodied" does not necessarily mean situated, in the feminist understanding of the term at least. The embodiment of cognition allows the mind to be understood physically and materially without reducing it to neurons alone. In the embodied mind theories I address, *embodiment* refers variously to the significance of emotion and feeling, the dependence of the brain on the rest of the body and its practical entanglements in the world, and the active character of perception and thought. This literature is understandably praised for contesting neuroreductionism and for opening up the brain-mind to the vagaries of the body. In naturalized philosophy, however, accounting for embodied cognition can entail a hunt for physical universals, on the one hand, or singular epistemic outcomes, on the other. To the question, *What difference do differences in bodies and embodiments make?*, the answer can effectively be *none at all*. I argue that this answer holds only for cognition that is generic, ideal, and normative. If, however, embodiment involves particular social locations, specific histories, differing personal and collective vulnerabilities, and varying and shifting capacities, then the outcomes of embodied cognition must be multiple. There is no singular embodied mind, then, but only embodied minds. Such heterogeneity, however, cannot be reduced to categories of the subject (or to discursive differences). Nor can it be described as residing solely in biology (reduced to morphological or brain differences). Drawing from Clark's extended cognition thesis, Haraway's idea of situated knowledge, and Garland-Thomson's notion of disability as mis/fitting, I argue for an assemblage model of the embodied mind to address the differences embodiment might conceivably make.

The idea that embodied minds differ and conflict also informs my reading of mirror neuron research, where the ability of neurons to mirror or simulate another's actions and emotions allows the brain to be innately social and intersubjective. On the view of embodied simulation theorists, this inherently social capacity is responsible for theory of mind and empathy. I discuss the dependence of this claim on the isomorphism of bodies and embodiments, which are understood as both uniform and unifying. To suggest otherwise, I refer to a famous, still painful example of racialized violence. If in some contexts, but not others, a set of dots can be perceived as a hand, what are the factors that allow a wallet to look like a gun? Or, more correctly, what are the factors that allow a particular body's actions to be read as imbued with one set of intentions (involving a wallet) rather than another (involving a gun)? My point is not that mirror neurons should be able to explain both readings and misreadings of mind—it is not all down to mirror neurons, in any case. Rather, my point is that perception takes place in worldly contexts that render automatic simulation a poor model for intersubjective understanding. Body-minds often perceive each other not simply as conspecifics, but as Others, whether racialized, gendered, sexualized, classed, or nationalized. These striations are generated within microinteractions as well as by macropolitical forces. How might they affect our responses to others—not only in intellectualist or symbolic but also deeply embodied and embrained ways? In other words, how might bodies and meanings participate in intersubjectivity in all its complexity, promise, and failure? Whether it is mirror neurons or some other neural mechanism at issue, the *social* in the biosocial brain must be more complexly understood.

My goal is not to close down discussions of biosociality, but rather to open them up. I have no wish to prove cultural over biological explanations, or to argue that social concerns trump material ones. In fact, I see little point in reasserting the distinctions between these realms. Returning to my discussion of kinship, it is undeniable in my view that evolutionary adaptationist accounts ignore both the queer predilections of nature and the diversity of kinship practices in human societies. But I find social constructionist accounts of kinship too empty; in denying material and fleshly realness to heteronormative kinships, they also evacuate queer ones. Neither approach comes close to addressing the bonds that do exist between beings with respect—that is, as both meaningful and materially real. These bonds comprise both historically specific meanings and structures, and materially specific transformations of embodied beings as they change each other. The ways (both welcome and not) that sustained bonds with others can leave their corporeal trace are, in fact, part of what is at stake in the social regulation of human kinships. To better grasp this, it is necessary to open up the complexity of culture to its neurobiological entanglements. These must be reclaimed from methods and models that disrespect the tangible diversity and specificity of lives.

NOTES

Introduction

1 "The Social Brain," RSA Action and Research Centre (Royal Society for the Encouragement of Arts, Manufacture and Commerce), http://www.thersa.org /action-research-centre/learning,-cognition-and-creativity/social-brain, accessed August 1, 2014.

2 An exception, it should be noted, was in the clinic, where neurologists chronicled the effects of brain lesions. Their tales of patients like Phineas Gage, a railroad worker in the nineteenth century whose frontal lobe injury reportedly left him with a radically transformed personality, were cited to illustrate brain localization, but they also gestured more broadly to the neurobiological underpinnings of the psyche. In recent years stories of Gage and other patients in the annals of neurology have been reprised through the explosion of neuroculture (Macmillan 2002; Sacks 1995).

3 I am referring here to the European Union–funded Human Brain Project, which has a budget of $1 billion, and to the National Institutes of Health's $100 million BRAIN Initiative. These initiatives were announced in 2013.

4 This is, Rose and Abi-Rached argue, specific to "advanced liberal democracies."

5 Rose (2006) argues that the individual is not reduced to her neurobiology, nor is she subjected to neuroscientific norms. Even as she is increasingly called on to think of herself in neural terms, she is not expected to conform to any singular conception of the brain or cognition. She is aligned with new, emergent brain ontologies made possible in part through biotechnological and pharmaceutical interventions, rather than with eugenic ideas of the normal and the pathological. For Rose, the openness of the neurobiological body to new and emergent forms of life releases us from the ethical and political worries of the eugenic era. This begs the question, though, of what new norms are invoked and what new pressures arise. What kind of brain—or brains—can or should one have? What models of neurocognition will be proposed and achieved? What conceptions of sociality? And how will these entangle with bodies, literally shape and reshape them (Mol 2002)?

6 See Gallese 2001, 2009, 2014; Iacoboni 2009; Iacoboni et al. 2005; and chapter 3, where I compare this paradigm with other ways of interpreting mirror neurons.

7 Although they do not always recognize their work in such terms, when they

identify the plasticity of the brain and its entanglement with social life, neuro-scientists and their interlocutors open up neurobiology to politics. Some, for example, depict normative social life (involving theory of mind, empathy, or kinship) as more or less a reflection of the brain's innately social character, raising the question of how to account for what they elide or erase—the conflicts, struggles, and inequalities in social relations. How do non-normative bodies, experiences, and relations fit into the biosociality neuroscientists are keen to describe? Others suggest that inequality literally shapes the brain, raising concerns over how to conceive neurobiological subjects in relation to power.

8 I thank an anonymous reviewer for succinctly stating this as a key contribution of this book, and Elizabeth Wilson (2004, 2010), who persuasively argues that neuroscientific data is not monolithic, and that it can be mined for feminist and queer insights.

9 On the incorporation of the brain into conceptions of the self, see, for example, Fullagar 2009; Martin 2010; Pickersgill and Van Keulen 2012; Rose 2007.

10 Many feminists writing about the neurosciences have made this case, including Fausto-Sterling 2012; Fine 2011; Joel 2014; and Jordan-Young 2010. Other citations can be found in chapter 1. While sex difference research has found new life in brain imaging technologies, it is rooted in mid-twentieth-century neuro-endocrinology that precedes the explosion of interest in neural plasticity and epigenetics. It relies on a conservative account of evolutionary biology, and it makes claims about gender roles, largely based on animal models, that have for decades been called into question by feminists. Although the resilience of this research program should not be underestimated, in its extreme biological determinism it could seem quaint in comparison to empirical and theoretical models of neural plasticity and biosociality. But feminist critiques of sexed brain research suggest the issue is not merely a lag in scientific progress. For a review, see chapter 1 and Fine et al. 2013; Hyde 2014; Schmitz and Hoeppner 2014.

11 Inasmuch as the centrality of brain science affects not just theory but praxis, not only intellectual life but also ordinary existence, it is no longer possible to dismiss the brain. It is not simply that references to the brain are ubiquitous; there is a lot at stake for many people, including "anyone implicated in discussions of morality, emotion, reason, intelligence, sanity, health, sexuality, personality and character." I argued this in an essay I called "The Neurocultures Manifesto" (Pitts-Taylor 2012a). A manifesto ought to be prescriptive, but how best to address these issues is not, in fact, as straightforward as a manifesto requires.

12 Feminist scientists have stressed the changes in the life sciences that make possible a more dynamic, developmental, and situated view of organisms. See Asberg and Birke 2010; Birke 1999; Oyama 2016; Thelan and Smith 2004.

13 As the conceptual boundaries between biological and social realms are eroding—as scientists are increasingly moving to address sociological concerns and social theorists are increasingly engaging with biological matter—the question is

whether key distinctions between what might be called "natural" and "social" approaches to the body (Pitts-Taylor 2008; Shilling 2003) remain.

14 This literature mostly draws from the middle work of Michel Foucault (e.g., 1975, 1979), and the early work of Judith Butler (1990, 1993). In this vein, my previous books explored how body practices reveal the body's semantic instability—its multiple and sometimes contradictory meanings. In my previous books I examine the cultural contests generated over the body and its modification (Pitts 2003; Pitts-Taylor 2007, 2008). In this vein, the specifically neurobiological body can be understood through the discourses and practices that infuse it with normative meanings, including gender and racial bias (Metzl 2003, 2011; Upchurch and Fojtová 2009) and neoliberalism (Martin 2000; Pitts-Taylor 2010; Rose 2007).

15 Although illness can be seen as saturated with symbolism, as Susan Sontag (2001) argued, and as a social construction, as many medical sociologists see it, there are many examples of feminists calling for realist approaches to illness and the ill body (Anzaldua 2002; Bost 2008; and Malabou 2012, to name a few).

16 I am thinking of a broad range of work in feminist and queer phenomenology and affect theory that takes up the experiential subject as well as the population (Ahmed 2006; Alcoff 2006; Berlant 2000; Butler 2010; Puar 2008; Saldanha 2006).

17 Disability studies scholars have criticized the medical model of the body for its individualizing and reductive account of disability, and its assumption that all bodies should meet normative ideals. Many disability scholars propose a social constructionist model that accounts for disability in terms of social barriers to full recognition and participation—biases, built constraints, stigmas, and ableism. However, some have argued for a "realist" attitude about the physical body and its capacities, which neither individualizes disability nor effaces the material, physical aspects of bodily variance. See also Lane 2009; Scully 2008; Siebers 2001.

18 In the 1990s Haraway (1991), Elizabeth Grosz (1994), Anne Fausto-Sterling (1992), Linda Birke (1999), and others argued against the separation of the social and the biological; their views are being affirmed in materialism.

19 I am ambivalent about this moniker, since in fact feminist theorizing of biology as entangled with the social isn't "new." But the term *materialism* on its own often suggests an uncritical engagement with the natural sciences, and the term *feminist materialism* is exclusive to feminist thought. The term *neo-materialism* can be used to describe contemporary modes of engagements with the natural sciences in both feminist and nonfeminist fields of inquiry, modes that stress the dynamism, flexibility, complexity, and plasticity of nature. See Alaimo and Hekman 2008; Cheah 1996; Colebrook 2000; Coole and Frost 2010; Davis 2009; Hird 2009; Kirby 2008, 2011; Stengers 2010; E. Wilson 2010.

20 For example, on physics, see Barad (2007); on dynamic systems biology see Oyama (2000a, 2000b); on epigenetics see Landecker (2011), Davis (2014) and

Weasel (in press 2016); on neuroscience see E. Wilson (2004, 2010, 2015) and Malabou (2008, 2012); on evolutionary theory see Grosz (2004, 2005).

21 With regard to the brain's immanent multiplicity, I am thinking especially of the work of Elizabeth Wilson (1998, 2004, 2010, 2015) and Catherine Malabou (2008, 2012). While Malabou takes a fairly uncritical approach to contemporary neuroscientific facts, Wilson does not; rather, she mines the paradoxes within neuroscientific thought to find resources for feminist theory.

22 These include theorists influenced by the work of Gilles Deleuze and Felix Guattari, such as Brian Massumi (2002), William Connolly (2002, 2011), John Protevi (2009, 2013), and Patricia Clough (2007).

23 See, for example, Claire Colebrook (2008) and William Connolly (2011).

24 Power's relation to the body is also being rethought. While power can be conceived as discursive, achieving its effects through representation, it is also material, achieved through the transformations of matter. For example, when technoscientific and biomedical practices refigure the body at the molecular level, engineer and patent gene sequences, harvest organs, or clone embryos, they not only denature human bodies but also extract biological life and capitalize it outside of the boundaries of the individual and the exclusively human. The treatment of biological matter as information, and the acceleration of data-driven mechanisms of governance and biocapital, suggests that power can work not only through discipline but also through securitization. Whereas the former shapes the subject by inscribing the body, the latter bypasses the individual subject in favor of generating, capitalizing, and managing statistical risk. Thus some theorists are less concerned with the intersectional subject and more with the production of populations through the informational and statistical management of life (Clough 2010; Clough et al. 2015; Foucault 2009). This emphasis requires attention to affect, understood as a "material intensity that emerges via the 'in-between' spaces of embodied encounters" (Pedwell and Whitehead 2012, 116).

25 See Omar Lizardo (2007, 2014); David Franks (2010); and Loïc Wacquant (2015). For a critical response see Pitts-Taylor (2012b, 2014, 2015).

26 Catherine Malabou (2008) simply declares herself a materialist in response to criticisms of the neurosciences. Doug Massey (2002) accuses sociologists of ignorance about biology; Lizardo (2014) makes a similar case. Rose and Abi-Rached (2013) argue for a "third way" in contrast to the antineuro detractors, who see neuroscience as biologically reductionist, and neuro evangelists, who treat it as a font of biosocial knowledge. For my arguments in sociology see Pitts-Taylor 2012b, 2013, 2014.

27 Among feminist materialists writing about the brain, some argue for new methodologies that could better respect the brain's diversity and dynamism (e.g., Joel 2014; Jordan-Young 2010; Rippon et al. 2014). They also argue for the brain's plastic and developmental character, its environmental embeddedness, and its situatedness in the whole, experiential body (Einstein 2012; Fausto-Sterling 2012; Fausto-Stirling et al. 2000; Van Anders and Watson 2006). They approach

neuroscience with epistemological and methodological reflexivity, see scientific observations as contingent and relational, and improve on simplistic, bifurcated categorizations of research subjects (Roy 2004). These efforts sometimes envision a "successor" science (Harding 1986), one that is less likely to do harm, is more helpful in bettering lives, and is put to the interests of social justice.

28 While uncritical empiricism reifies brain facts as unmediated reflections of reality, Bruno Latour laments that critique is too often limited to describing the conditions under which the facts of biology are known or generated. As such, it is only negative. Neither approach does enough to address the stakes of brain knowledge. Siebers (2008) embraces a critical realism; I take up Barad's notion of agential realism in chapter 1.

29 Biology's ongoing multiplicity—its refusal to be isomorphic or self-identical—need not be understood as emanating from exclusively cultural influences, nor treated as evidence for human exceptionalism. Rather, it may reflect the ontological heterogeneity and diversity of nature (Grosz 2004; Hird 2009; Kirby 2008; Oyama 2016), the promiscuity and multiplicity of natural kinds (Rouse 1998), the probabilistic rather than predictable character of causalities (Rouse 1998), and the generally "complex, messy and richly various" character of biological life (Medawar 1969, 1, quoted in Einstein 2012). Yet how to address the neurobiology's specific malleability and its imbrications in experience is a daunting challenge.

30 For a discussion of the brain as performative, see Dussauge and Kaiser (2012a, 2012b) and Kaiser (in press, 2016). I address the distinction between discursively and materially performative conceptions of the brain in chapter 1.

31 Given my use of disability studies to conceptualize cognitive difference, I want to clarify how my argument relates to, and differs from, arguments for neurodiversity. *Neurodiversity* describes the view, often associated with the Autism Pride movement, that genetic variations create different kinds of brains and neurocognitive styles (Ortega 2009). Echoing the social model of disability, advocates for neurodiversity argue for the positive contributions of diverse cognitive styles and reject their pathologization and stigmatization (Armstrong 2010). They claim that the disadvantages of these variances, as manifest in autism and other conditions, are primarily rooted in social norms and barriers to participation created by "neurotypicals," rather than in disabilities that are inherent to diagnosable conditions. The movement highlights the need for greater regard for and sensitivity to neurocognitive variation. It also sheds a critical light on the use of autism as a model for almost everything, it seems, that could go wrong with cognition or emotion. (To cite just one example, Malcolm Gladwell [2005] uses the phrase *temporarily autistic* to explain theory of mind failures.) However, my argument for embodied multiplicity is not reducible to neural difference, but encompasses the broader notion of complex and situated embodiment. I argue that cognitive and affective dissonance can be related to social stratifications such as race, class, gender, and disability, as well as physiological variation and differen-

tial relationships with the built world. Further, my discussion calls into question the "neurotypical" and its reification. Not only is autism cited to rationalize a broad range of neuroscientific research programs, but it is also used to shore up claims about what constitutes typical traits, relations, and experiences, including those related to sex/gender (Gillis-Buck and Richardson 2014). The concept of neurodiversity on its own does not do enough to unsettle these claims.

32 The divisions between empirical and theoretical research programs are not strict; rather, neuroscientists rely heavily on frameworks from other fields to identify research problems, to make sense of findings, and to argue for their relevance. The embodied simulation theory of mirror neurons is a good example; the significance of mirror neurons was established initially through interventions in debates about theory of mind taking place in philosophy of mind and psychology (Gallese and Goldman 1998). The framework of simulation has shaped subsequent discussions about mirror neurons in relation to empathy (Gallese 2001, 2014), concepts and metaphors (Gallese and Lakoff 2005), art and aesthetics (Freedberg and Gallese 2007), and sociocognitive disorders (Iacoboni 2008; Gallese et al. 2009). The strongest claims about mirror neurons depend on the idea of a shared manifold—a common phenomenological experience, motor schema, and relation to objects—that unites humans as conspecifics through their embodied relations in the world.

Chapter 1. The Phenomenon of Brain Plasticity

1 James dedicated a chapter of his classic *Principles of Psychology*, published in 1890, to considering the question How and why do we gain continuity in our thoughts and behaviors over time? In animals, he wrote, habits are given by instinct, whereas in people they are generally considered acts of reason, products of disembodied, conscious thought and deliberation. A pragmatist, James rejected both instinct and reason as explanations for human habit. James recognized plasticity as a material property that has its own force, outside of human control, but like his contemporary Ramon y Cajal, he was also concerned with its personal and social implications. We should guard against bad habits "as the plague," James wrote, and we can capitalize on the inculcation of good habits.

2 Brain tissue looked much different than other biological tissue under the microscopes of the nineteenth century, enough so that its cellular composition was in question. In 1873 the Italian neuroanatomist and histologist Camillo Golgi, working in a lab he set up in the kitchen of his apartment, used a new stain, silver nitrate, to look at nervous tissue. He saw cell-like structures that looked different from other cells, having two different kinds of extensions or "processes" coming out of them (dendrites and axons). These were densely intertwined and formed, he thought, a holistically functioning network of electric nervous impulses. In 1887 Cajal, a Spanish army doctor, medical illustrator, and anatomist, also working out of his kitchen, borrowed Golgi's method. Using the same technology,

looking at similar tissue, he saw something different (DeFelipe 2006; Shepard 1991).

3 For example, in 1890 Weidersheim claimed to observe neural plasticity in his study of *Leptodora hyaline*, a crustacean that lived in deep, freshwater lakes in northern Europe. The crustacean was unusual in being "absolutely transparent, almost invisible in a glass of water," as described in *Nature* in 1897, cited in DeFelipe 2006). Able to see right into to its living organs, he claimed that cells in its cerebral ganglion continuously changed their shape in a "slow and flowing" manner (Weidersheim 1890, cited in DeFelipe 2006). Cajal's student Tanzi hypothesized that such changes in the ganglia resulted from changes in neurons' connections with other neurons and, further, that the strengthening of connections is linked to the consolidation of memories and the learning of motor skills (DeFelipe 2002, 2006).

4 Pharmacologists and neurophysiologists vigorously debated the mechanisms of synaptic transmission for decades; only in the 1950s did they agree on Henry Dale's chemical explanation involving neurotransmitters, which is still understood to explain most neural communication. Dale, along with Otto Loewi, won the Nobel Prize in 1936 for their work on acetylcholine as a neurotransmitter, but many neurophysiologists (most notably John Eccles) rejected chemical transmission in the central nervous system in favor of an exclusively electrical model of neural communication. In chemical transmission an electrical charge in the axon of a neuron, or an "action potential," stimulates the release of a chemical (such as acetylcholine or serotonin) from a vesicle in the cell. It acts as a neurotransmitter, traveling across the synapse and binding to a receptor on another neuron's dendrite. This binding can stimulate an excitatory electrical charge, which increases the likelihood of the second neuron generating its own action potential and transmitting its signal to yet another cell. (It can also generate an inhibitory charge, which decreases such likelihood.) In this way chains of electrical connection are created. Beyond neurotransmitters, other neuromodulators (such as hormones) have also been recognized as influencing synaptic transmission, and all of these chemicals are thought to influence each other as well. Loewi found evidence for chemical neurotransmission in his famous frog study, the design of which came to him in a series of dreams. He removed the beating hearts of two frogs in vivo, one with the vagus nerve still attached. He changed its rate with electrical stimulation of the nerve. He was able to change the rate of the other by applying fluid from the first heart. The neurotransmitting chemical in this fluid was eventually identified as acetylcholine, which Dale had identified earlier as a neurotransmitter. The chemical explanation for synaptic transmission competed with John Eccles's insistence that transmission is primarily electrical. The Dale-Eccles debate was highly contentious, reportedly involving public rows (as did the Golgi-Cajal debate), and it was not until the very end of the 1940s that Eccles embraced Dale's position. Until the 1950s, as Dale wrote in notably sexist terms, chemical transmission was treated "like a lady with whom the neurophysiologist

was willing to live with and consort in private, but with whom he was reluctant to be seen in public" (cited in Valenstein 2005, 131). Dale was ultimately victorious; chemical transmission is now understood as the primary form of communication in the nervous system. The current view is that chemical synaptic transmission characterizes the majority of communication in the brain. In 1950 gap junctions, which seem to allow electrical current to pass directly from one cell to another, were also identified.

5 Further, differences in intervals between pulses had an impact on how many synapses were affected. In the 1980s Hebbian plasticity and long-term potentiation helped to build a connectionist model of cognition. Connectionists argue that cognition comprises the activations of networks of neurons that are distributed across the brain. Neurons at different locations in the network calculate the strength of their own activation based on the firing rates of inputs from other neurons. These calculations also are weighted by past exposure to stimulation, which affects the neurons' amplitudes. The knowledge held in neural networks "is contained in and defined by its very architecture, in the connection weights that currently hold among all units as a function of prior learning" (Bates and Elman 2002, 431).

6 In *A Thousand Plateaus*, Deleuze and Guattari characterize synaptic plasticity in terms of the agency it affords to neurobiological matter: "What are wrongly called 'dendrites' do not assure the connections of neurons in a continuous fabric. The discontinuity between cells, the role of the axons, the functioning of the synapses, the existence of synaptic microfissures, the leap each message makes across these fissures, make the brain a multiplicity immersed in its plane of consistency or neuroglia, a whole uncertain, probabilistic system . . ." (1987, 15). See Murphie 2010 and Dewsbury 2011 for discussion.

7 Some models argue for considerable experience dependence in the brain's maturation. In their manifesto of neural constructivism, Steven Quartz and Terrance Sejnowski (1997) argue that humans begin life with a "protocortex," which has basic circuitry but develops regional specializations only through input from the rest of the body and the environment to the brain. They argue that cortical wiring is progressively elaborated through the interaction of mechanisms that are dependent on experience with developmental trajectories that are intrinsic (Quartz 1999). Although neural constructivists acknowledge genetic processes, they emphasize epigenetics, or the varying ways genes are expressed as the environment affects them.

8 Neuroimaging technologies are used to create visual images of brain structures and processes. MRI scans, for example, create computerized, two-dimensional images of brain tissue that look similar to X-rays, whereas fMRI scans depict oxygenation levels in the brain to approximate neural activity in different brain regions. There are a number of methodological limitations to imaging research that are addressed in the growing critical literature on neuroscience. Joe Dumit (2004), for example, has offered an exhaustive account of the fallacy of "picturing

personhood" through brain scans, which do not photograph or film the brain in action but rather statistically craft images of brain activity based on relative levels of activation in different areas. Producing an fMRI image involves many steps of calculation, each of which involves a set of decisions, such as choosing statistical thresholds, which influence the result. Further, deciding what those images mean is a matter of interpretation. Scans often are read through reverse inference, where particular cognitive processes are inferred from observed activations in areas thought to be associated with such processes. Cordelia Fine (2011) addresses other methodological problems with brain scan research measuring sex differences, including consistently small sample sizes, the difficulty of controlling for bodily variables such as breathing rate and caffeine intake, and controversies over statistical procedures.

9 For example, the corpus callosum, a bundle of nerve fibers that connect the two hemispheres of the brain, and the cerebellum, located at the back of the brain, are considered to have distinct trajectories. The former is thought to develop "structurally throughout life, but most dramatically during childhood and adolescence" (Luders et al. 2010, 1098S), whereas the latter seems to be plastic much longer, developing into the early twenties (Giedd 2004). The purportedly less plastic corpus callosum has been associated in the literature with problem solving, compared with the more plastic cerebellum, which has long been linked to motor coordination and balance. The cerebellum's role has more recently expanded to include the coordination of cognitive as well as physical tasks.

10 For example, if the Supreme Court considers this knowledge in deciding whether the execution of teenagers is unjust, as in *Roper v. Simmons* 2005.

11 Based on their review of the imaging literature that detects different task-related patterns of activity in the brains of research subjects grouped by SES, as well as their own experiments that use a battery of tests on children of low- and middle-income backgrounds, they propose that SES affects primarily two of the five (Noble et al. 2005) or seven (Farah, Shera, et al. 2006) brain systems they identified.

12 Edward Taub's controversial work offered surprising evidence of major cortical remapping. To study the relation between sensory input and motor use, Taub deafferented the limbs of rhesus and macaque monkeys to permanently cease the feeling of stimuli in their arms. In the wake of a campaign by animal rights activists, the monkeys were removed from Taub's lab. Over a decade later, when the monkeys were euthanized, a group of researchers examined their somatosensory cortices. They found that up to 14 mm of cortex that likely would have received stimuli from the arms was instead receiving stimuli from the face (Pons et al. 1991). This was more extensive cortical remapping than ever documented before, leading Taub to hypothesize that the plasticity involved not just the axons of neurons in the cortex but the thalamus or brain stem, which would indicate a more global mechanism of plasticity (Clifford 1999).

13 The researchers also experimented with using cognitive reinforcements to boost

this topographic malleability, finding that attention to a task is a crucial factor in transformations of the cortex (Merzenich 2012).

14 In lesion studies, which correlate structures with functions by examining the behavioral impact of damage to particular brain areas, the hippocampus has long been linked to spatial memory and navigation.

15 Their 2006 study compared taxi drivers with bus drivers (who, unlike taxi drivers, follow planned routes and thus do not have to memorize a huge amount of spatial information). In this study taxi drivers also measured comparatively more gray matter volume in mid-posterior hippocampi and less volume in anterior hippocampi. Interestingly, however, in a battery of tests, their performance on tasks in acquiring *new* spatial information was poorer than that of the bus drivers. To test this further, they took scans of seventy-nine trainees (and thirty-one controls) over the three to four years they took to prepare for the examinations. They found no differences in the structures of two regions of the hippocampus at the start; later, however, they measured greater volume in the posterior hippocampus in trainees who passed the exam, but not in the controls or the trainees who failed the exam (Maguire et al. 2011).

16 In the 2011 study Maguire et al. took scans of the hippocampi of seventy-nine trainees (and thirty-one controls) over the three to four years they took to prepare for the tests. They ended up with three groups: those who did not study at all (since they were controls, not aiming to be taxi drivers), those who studied and failed, and those who studied and passed. They found no differences in the structures of the hippocampus between any of the groups at the start. Later, however, they measured greater volume in the posterior hippocampus in trainees who passed the exam, but not in the controls or the trainees who failed the exam.

17 A 2011 review paper published in *Frontiers in Neuroendocrinology*, for example, asserts that gender identity and sexual orientation are "programmed into our brain during early [prenatal] development. There is no evidence that one's postnatal social environment plays a crucial role" (Bao and Schwabb 2011, 214). Depending on whom you ask, male and female brains are very different, very similar, or a little bit different. Anne Fausto-Sterling (2012) warns that many of the reported differences, such as percentage of white matter versus gray matter and relative size of the corpus callosum, hippocampus, and amygdala, disappear when studies control for the relative size of the brain in men and women. (Because of their larger body size, men's brains are 10% bigger on average.) And Cordelia Fine (2011) and Rebecca Jordan-Young (2010) caution that isolated reports finding sex differences in the brain need to be assessed in conjunction with others; many differences, such as the degree of lateralization across the right and left hemispheres, disappear when examined across multiple studies.

18 Although the function of the straight gyrus is not settled in neuroscience, they cite several fMRI studies that report activity in this area during social intelligence tasks, as well as lesion studies that show deficits in social intelligence in patients with brain damage in this area.

19 Feminists have offered extensive critiques of brain organization theory, address-
    ing methodological flaws, interpretive problems, and ideological assumptions. In
    the most extensive and systematic critique, Jordan-Young (2010) conducted a
    meta-analysis of all of the studies published in the first thirty years of brain orga-
    nization theory. She found that sex and gender are inconsistently operationalized
    and measured across studies testing the theory. There is no standardized method
    of defining feminine and masculine behavior, homosexual and heterosexual
    orientation, and, in ambiguous or intersex cases, even male and female. The cat-
    egories mean something different from one study to the next. They seem to be
    applied arbitrarily, but Jordan-Young argues that the categories reflect research-
    ers' heteronormative biases. She also describes how the studies rely on "quasi-
    experiments" due to the limitations of studying brain hormones in utero. She
    argues that this research should not be taken as seriously as it is because of its
    lack of direct evidence and because of the inconsistent design and results across
    studies (Jordan-Young 2010). Researchers commonly assume that both biological
    and behavioral changes observed in animal and human studies will last beyond
    the period of study and even throughout the lifespan. In other words, they as-
    sume one of the very points they seek to prove—that observed differences are
    inborn and fixed. Further, they use brain organization theory to explain that
    differences that could be interpreted as the result of socialization (Jordan-Young
    and Rumiati 2012). Jordan-Young (2010) proposes a research program that bet-
    ter captures the plastic nature of the brain. Such neuroscience would focus on
    "the dynamism of development—understanding processes more than perma-
    nent states, development rather than 'essences'" (291). This requires assuming
    that physiology is iterative, so that "the current state of the organism interacts
    with each [experiential] input" (286). Instead of tracing a linear causality, the
    research would try to capture the ongoing feedback and exchange between all
    of the factors that might be hypothesized to influence behavior, and to explore
    the stability of these influences over time. Jordan-Young offers an example of
    how animal studies that test the effects of prenatal hormones on sexual behavior
    could recognize plasticity: "Such experiments would require testing hormones-
    during-critical-periods as the *first* important variation in environment, and the
    phenotype that results from early hormones as an *interim state* in the organism's
    development. Next, it would require systematically testing how early hormone
    exposures affect the behavior of animals who are reared in different ways, sexu-
    ally socialized in different ways, exposed to different subsequent hormone envi-
    ronments, have different diet and exercise regimens, and so on" (289).

20 The research by Wood et al. has been highlighted as an example of how sex/
    gender in the brain cannot be neatly attributed to the different socialization of
    females and males (see Eliot 2009), but it raises as many questions as it answers.
    Wood, Heitmiller, et al. compared not only the straight gyrus of the subjects
    by biological sex, but also by the subjects' gender, as measured by a standard
    psychological index of masculinity and femininity addressing interests, per-

sonality type, and abilities. When comparing by sex, women had larger straight gyri. However, a different picture emerged when comparing by gender. The researchers found that regardless of the sex of the subject, more feminine scores correlated with higher scores on social intelligence tests, *as well as* larger straight gyri. For example, males who were judged to be more feminine had a larger straight gyrus than the male average, as well as better than the average male scores on social intelligence. Wood et al. see this as evidence that social intelligence is highly gendered (not just sexed), and that gender is linked to the size of this strip of cortex. Despite having, in theory, the same etiology in prenatal hormones, Wood, Heitmiller, et al. suggest masculinity/femininity in the brain does not perfectly overlap with the dimorphism of the genitals. Bodies with male reproductive organs can have more, *or less*, "male" brains. While in their view gender remains an effect of biology, this research also demonstrates the variability of gender patterns, as well as the variability of brain morphology, *within* sexes. The study is being cited as a welcome challenge to the idea that "brain differences are a simple product of the Y chromosome" (Eliot 2009, n.p.; Kaiser forthcoming). But while it pulls apart the conjoined categories of sex/gender, it is also consonant with a long-standing effort by sex difference researchers to locate bifurcated, universal categories of gender (masculinity and femininity) in the brain and to position the brain as more significant than the genitals in the development of sexual identity and orientation (Kraus 2012). Here, *gender is not conceived as plastic*; instead, like reproductive organs, it is determined by prenatal brain development.

21 This possibility is interesting in light of the follow-up study by Wood, Murko, et al. (2008), which measured social intelligence and the straight gyrus in seventy-four children between the ages of seven and seventeen. They found that girls, like women, did better on social intelligence tests than their male counterparts. But instead of finding larger straight gyri in girls than in boys (what they expected after finding this pattern in adults), they found the opposite—the boys had proportionally larger straight gyri than girls. The relative difference of this area, it seems, *reverses* over the lifespan. This surprising finding did not lead Wood, Murko, et al. to question their larger hypothesis or its evolutionary underpinning. Instead, they explain this finding by referencing the earlier maturation of girls' brains: "for this region females may be ahead on the developmental trajectory, with earlier gray matter pruning. By adult years, boys catch up and complete developmental gray matter changes and straight gyrus size ends smaller than females." This explanation, however, does not address why males' developmental delay is more pronounced in this region than others. Nor does it explain why male brains in this area have not caught up by adulthood. We are left with a fuzzy proposition: In girls, a smaller straight gyrus is linked to greater social intelligence in youth, and bigger straight gyrus size accounts for their greater social intelligence when they are older.

22 The concept of performativity derived from Butler's work (1990, 1993, 2004) describes how culture shapes embodied experience and creates an historicized

body-subject. Performativity is a productive process; it genders the subject and sexes the body according to pervading norms. It comprises the repetition of an "act that has already been rehearsed" (Butler 1990, 272). In contrast to the idea of socialization, however, performativity resists any temporal stratification or sense of completion. While it is embodied, it is not tied to a developmental context, nor stratified by age or stage of maturity; instead, gender performativity requires constant reiteration over time. Butler understands biological matter not as a static ground of experience, but rather as engaged in a process of materialization, one that sediments over time to give the impression of originary stasis. For Barad, however, Butler's conceptualization of performativity "ultimately reinscribes matter as a passive product of discursive practices rather than an active agent participating in the very process of materialization" (2007, 151). It is culture or discourse that creates the script for the body, and it is culture that, in the guise of nature, gets performed. To be treated as an active agent, the body has to be recognized in its material specificity. Following insights from quantum physics, Barad insists that matter is agential; it is always in the process of its own making. Thus all matter, including biological matter, can be characterized in terms of an event, a doing, "a congealing of agency" (Barad 2007), or a becoming. How does biological matter gain its boundaries in relation with the rest of the world? How, when, and under what conditions does it sediment or congeal? How, when, and where does matter come to matter?

23 Mani et al. (2013) conducted two experiments to see how poverty might be considered a circumstance that affects cognitive performance. In the first study they prompted subjects to think about finances; in those with low SES this impeded test performance, whereas it had no effect on those with high SES. In one study they tested a group of farmers before and after harvest. The farmers performed better on cognitive tests after harvest, when they were presumably less financially vulnerable, than before. Mani et al. argue that financial stress affects the subject not merely indirectly through well-known biological markers but also through cognitive routes: Poverty itself takes up neural resources. "The poor, in this view, are less capable not because of inherent traits, but because the very context of poverty imposes load and impedes cognitive capacity. The findings, in other words, are not about poor people, but about any people who find themselves poor" (980).

24 This includes Farah, Shera, et al. 2006; Farah, Noble, et al. 2006; Hackman and Farah 2009; Hackman et al. 2010; Noble et al. 2005.

25 Similarly, Cajal was interested in the prospects of improving the brain through mental exercise. Because he believed that plasticity closed down in early life, except in cases of injury, he wrote, "it is a cultural task of the society to shorten the time required by the brain cells to reach their perfection" (Berlucci and Butchel 2009, 313).

26 This quote is from William James, "Bergson and Intellectualism," Lecture VI, in *A Pluralistic Universe*, cited in Rubin (2013).

Chapter 2. What Difference Does the Body Make?

1 The materialist view of mind was made possible in part by the convergence of scientific fields that had previously separated questions of the brain and mind, and by a new model of cognition that was not limited to computers but could be applied to the organic brain. Cognitive science worked on cognition by developing formal computational models rather than the brain itself, and neuroscience explored the brain without dealing with questions of mind. The connectionist view of cognition empowered both to take a stand on mind and consciousness. Rather than the abstract processing of symbols, connectionism explains cognition as the operation of neural networks distributed across the brain. In the last decade of the twentieth century, cognitive science became neurocognitive science; philosophy, Churchland hoped, would become neurophilosophy. While they once eschewed tackling the mind, "[f]ew neuroscientists now take a non-naturalist position, and still fewer hold to a principled agnosticism on the mind-brain question. The vast majority believe in physical realism and in the general idea that no nonphysical agent in the universe controls or is controlled by brains. Things mental, indeed minds, are emergent properties of brains" (Mountcastle 2000, 1). Philosophers could explore the mind not as the product of disembodied, abstract reason, but rather of physical, biological processes that are (in theory) measurable and locatable.

2 Ruth Leys explains: "since somatic markers signal the mixing of innate and learned components of our affective responses, the somatic marker hypothesis suggests a mechanism for conceptualizing how culture and the body interact. Somatic markers are thus said to be culturally influenced 'gut reactions' that provide guidelines for decision making" (2011, 464).

3 In their early work, *Metaphors We Live By* (1980), Lakoff and Johnson advanced the argument that metaphors, which they see as the primary conceptual building blocks for human reason, are informed by embodied experience. Twenty years later, in *Philosophy in the Flesh* (1999), they gave this argument a neurobiological foundation.

4 For example, Haraway noted that the world appears different when viewing it through a dog's eyes, a camera, a spy satellite, a magnetic resonance imaging machine, or any of the countless technological mediations used to visualize the world. Vision, she contends, is not a matter of neutrally observing an objective world that has a fixed and indisputable meaning. Nor is it a matter of passively perceiving it from a stable vantage point. Rather, vision is an active, productive process involving an assemblage of capacities, from the neural and other bodily systems of organisms to the "prosthetic technologies interfaced with our biological eyes and brains," that make perception possible (Haraway 1988, 589).

5 Clark's theory has led to consideration of how cognitive economies are assembled. For example, they may be "softly-assembled," that is, temporary and task-specific. Their organization may be interaction-dependent, in that their parts do

not have rigidly fixed roles but rather fluid ones shaped by the dynamics of all of the parts in the assemblage (M. Anderson et al. 2012, 3). A dynamic cognitive economy might be constituted by novel or repurposed configurations of neural networks, body parts, tools, and objects in the environment. Dynamic cognitive economies include other people, as when body-subjects achieve "plural" subjectivity through interpersonally coordinated tasks or group cognition (9). Multiperson cognitive systems, of course, raise considerable challenges to the assumptions of classical cognitive theory and Cartesianism more generally. We might gather from the extended view that "the information flow between mind and world is so dense and continuous that . . . the mind alone is not a meaningful unit of analysis" (M. Wilson 2002, 626). Dynamic systems theory takes this position, seeing constant, reciprocal, and productive flow between brain, body, and the surrounding world (Clark 2001, 128). The breakdown of cognitive boundaries and the dismantling of an internal cache of representations may suggest that there is a wholly post-Cartesian subject. Clark himself rejects such a view. While affording maximal plasticity in cognitive processes, the anti-internalist view does not well explain, according to Clark, "our puzzling capacity to go beyond tightly coupled agent-world interactions and to coordinate our activities and choices with the distal, the possible and the non-existent" (1998, 44). In other words, it does not account for the capacity of an individual to daydream, project into the future, hypothesize, or otherwise detach thought from actual circumstances. He favors instead what he calls a minimal Cartesianism, where *most* cognition can only be understood in the context of interaction in the external world, but an internal world of representational resources nonetheless exists; the latter, however, are tied to language and other external resources.

6 As Joseph Rouse (2009) has pointed out, Smith's standpoint theory and Haraway's notion of situated knowledge have in common a naturalistic orientation that prioritizes experience over rational and abstract reason. They embrace a phenomenological view of mind as situated in the "active and perceptual world" (202). Both theories adjudicate knowledge claims not through internal epistemic standards, but rather external ones. Haraway embraces a practical and critical realism, whereas standpoint theory holds that knowledge claims arise from lived experience. In both cases knowledge claims are judged "by what they enable us to see, say, and do, not the other way around" (201), and the whole, feminist epistemologies "situate knowledge and epistemic warrant within the world, amid our interactions with other agents, rather than in an abstracted space of representations" (201).

7 The term *situated* here refers to the embeddedness of "representations of the world, learning, memory, planning, action and linguistic meaning in the body's environment, conceptual structures, tools and social arrangements" (Solomon 2007, 413).

8 This insight came to Smith partly through her participation in feminist consciousness-raising groups. For Smith, "The Cartesian subject escapes the

body, hence escaping the limitations of the local historical particularities of time, place, and relationship. When we began with our experiences as women, however, we were always returning to ourselves and to each other as subjects in our bodies" (1992, 89).

9 African American women, for example, have historically different experiences than both white women and black men, and Patricia Hill Collins (1990, 2000) argues that their collective knowledge offers a distinct epistemology. The concept of intersectionality (Crenshaw 1991) resists essentializing body-subjects based on sex/gender, and it allows for the inclusion of many possible configurations of subjectivity. For these reasons, it has been the preferred framework for acknowledging the differentiating effects of social relations and stratifications on individuals and groups.

10 The embodied character of knowledge contests assumptions of rationalism and scientific objectivity. If rationalism assumed a male subject, mainstream phenomenology assumed a generic (male) body. Smith (1988) drew from Alfred Schutz's social phenomenology to situate embodiment in an institutional context. Men could partake in abstract rationality, the dominant logic of the public sphere, by relying on women's labor to take care of bodily needs, while their location in the domestic sphere meant that women had no choice but to attend to the everyday, actual, practical constraints of living. Nancy Hartsock (1983) argued from a Marxist-feminist perspective that men and women are differently positioned in relations of ruling through the gendered division of labor, which gives them divergent experiences and vantage points. Iris Marion Young (1990) argued that bodily engagement with the world enacts social differences, and thus there is no generic, ungendered phenomenological body. Later feminist phenomenologists address the perceptual and embodied differences that not only gender but also race and sexuality make (Ahmed 2006; Alcoff 2006).

11 The intersectional approach still involves the "homogenizing generalizations that go with the territory of classification and categorization" (McCall 2005, 1783). Intersectionality theory has the effect of securing body-subjects in stable (even if intersected and multiple) social locations, which tend to be treated as both sites and sources of identity. Critics challenge this framework in two respects. First, they contest the assumption that the body provides a stable location for experience. If the body is not passively inscribed by culture or securely located by social relations, but rather is always materializing, the body does not provide a ground for identity or subjectivity. Relatedly, they contest the reification of the subject, which depends on "an epistemological capture of an ontologically irreducible becoming" (Puar 2012, 54).

12 For example, in defending her work against criticisms of essentialism, Smith (1992) insisted that embodied knowledge is not reducible to the subject. Instead, she argued for an ontology of the actual, where the *actual* is what is prior to the distinction of the subject and object. If the subject and object do not arrive already delineated, but rather gain their specificity only within experience, then

epistemic inquiry cannot begin from differences in subjectivity; it must instead examine the conditions that generate difference. Relatedly, Puar (2012) notes that in Crenshaw's original depiction of intersectionality, the emphasis was not on the stability of an interlocking grid provided by race/gender/class, but rather on the movement of persons between multiple positions.

13 One of the risks of dispensing with the intersectional subject altogether, Puar says, is overlooking that for "some bodies—we can call them statistical outliers, or those consigned to premature death, or those once formerly considered useless bodies or bodies of excess—discipline and punish may well still be a primary apparatus of power" (2012, 63). In other words, for Puar, intersectional subjects still exist to the degree that disciplinary practices do. But the issue at hand is not simply whether, and for whom, power works discursively or affectively (if in fact they should be strictly distinguished). It is also whether, how, and to what degree experience sticks or congeals, and how it does so differentially. The task is to address embodied beings who both feel and think, and whose experiences substantially affect them, without reducing them to discursive constructions, on the one hand, or free-floating individuals, on the other. How might such beings be addressed? If the term *subject* implies a discursive construction, the more phenomenological term *body-subject* insists on a being's material reality. But *body-subject* also seems to presuppose stability. However, a body-subject can be thought of as an agentic assemblage, whose assembling is not generic and universal but specific and multiple. Lisa Blackman, for example, adopts a pragmatic conception of subjectivity to address the conditions that allow it to manage some sense of what William James called its "singularity in the face of multiplicity" (2008, 138). She sees the subject as "neither fully open nor closed" (2012, 189). Thus, body-subjects actively hold together; they are distinctly situated in space and time, while also potentially distributed across these, joined to other bodies and objects through biotechnological practices, affective forces, and communication technologies, and open to further configuring and refiguring. They also hold traces of experience that in some way provide a condition for future experience.

14 An assemblage is a combination of parts that can be refigured, or refigure themselves, into other combinations, as opposed to an organismic whole whose parts cannot be rearranged. Applied to the concerns of intersectionality, assemblage theory does not map the body-subject onto vantage points rooted in socially determined locations, or embodied identities that are stable. Rather, it examines the movement and rearrangement of elements that, in specific temporal and spatial contexts, constitute the racialized, gendered, or sexualized body. Assemblage theories can help conceptualize the embodied mind as a *becoming* that is constituted only through its interactions in the world. As agentic assemblages, embodied minds are not essential, fixed, or universal. Rather, they have a relational ontology, shaped in and through experience with other body-minds, objects, and entities in the world.

15 See Price and Shildrick 2002 for a related discussion of touch.

16 This affirms embodied realism while putting pressure on it to make room for epistemic difference. If common bodies create common knowledge, unusual bodies create marginal but (Scully would argue) no less valid knowledge. This could mean something like epistemic pluralism that includes *non-pathological* "disability knowledge." To make this case, however, Scully explicitly brackets any discussion of cognitive or mental disabilities. But a broader sense of bodily variance could include brain variance, and thus, neurodiversity.

17 Given the concerns flagged by Martin, many readers might consider this fortunate. The denial of epistemic difference, however, is objectionable on feminist grounds, and it is also problematic from the point of view of disability studies.

18 This view disrupts a sense of disability as the property of any individual, and therefore as a basis for fixed identity. Instead, disabled experience is a "process of fluid encounter" between bodies and worlds, including other bodies (Price and Shildrick 2002, 64).

Chapter 3. I Feel Your Pain

1 Methodological and interpretive critiques of mirror neuron research are extensive. For example, there is disagreement on whether and how much mirror neuron activity is actually discernible in human brains (Welberg 2010). There is a lack of agreement about what mirror neurons do in humans and what kinds of actions are actually mirrored. Mirror neurons in humans are located in brain areas thought to be significant for language, yet the mechanisms by which language and memory work with mirror neurons are far from clear (Damasio and Meyer 2008; Mahon and Carmazza 2008). Some critics have also pointed out that other kinds of neurons are capable of higher-level representations addressing goals, intentions, categories, and concepts (Hickok 2009), or, as I discuss, questioned the idea of simulation as achieving empathy or intersubjectivity (Wahman 2008; Jacob 2009). And compared with other multisensory neurons, "very little is known" about their development and their individual and networked computations (Heyes 2010a, 789).

2 Malcolm Gladwell (2005) also discusses the Diallo case as an example of theory of mind failure. Rather than attributing the case to racism, as many critics do, he attributes the mistakes of the four police officers to what he calls "temporary autism" caused by haste, stress, and physiological arousal. The use of autism as a general term for cognitive (or precognitive) theory of mind failure is questionable, made no less so by the fact that mirror neuron researchers are trying to explain autism through mirror neuron systems, as I discuss. Gladwell's discussion of race is also troubling. Having dispensed with the explanation that the officers were overt racists, he writes: "They [the officers] are in the line of fire, and what Carroll sees is Diallo's hand and the tip of something black. As it happens, it is a wallet. But Diallo is black, it's late, and it's the South Bronx, and under those circumstances we know that wallets invariably look like guns" (243). Gladwell

points to, but does not explore, the relationship between mind-blindness and race. Later I probe mirror neuron theories to question why, and *to whom*, wallets held by black men could "invariably" look like guns in the South Bronx. I argue for investigating how preconscious as well as propositional cognitive processes can be influenced by racism.

3 Contagious shooting is a highly contested phenomenon; there seems to be little empirical evidence to suggest that police officers are more likely to fire, and fire more bullets, because other police officers have done so. But police departments treat contagious shooting as a reality because police officers themselves have described such events, and they have also been observed in police training sessions (Rostker et al. 2008).

4 Recommendations to combat contagious shooting include that police officers not only memorize policies about deadly force but also be asked to practice their decision-making skills regarding the use of deadly force in situations as close to real life as possible. They also are encouraged to undergo target identification training, in which "we must make shooting a no-shoot target unacceptable behavior" (Joyner 2008, n.p.). Since the empirical existence of contagious shooting is so contested, it is difficult to say whether any of these recommendations make a difference.

5 Reportedly, the other officers did not merely hear or see Officer Carroll firing his gun; they also heard him yell "Gun!" to indicate that the suspect was armed; therefore they had explicit information to contend with as well (Fritsch 2000).

6 There is, as Maria Brincker puts it, an "intentional involvement with the world already in our perceptual pre-action relation to it—and it is this involvement along with background considerations that inform and shape our actual action choice (2011, 291).

7 In addition, there are also hypotheses about the relevance of mirroring for more complex aspects of cognition, such as aesthetic appreciation and the generation of concepts. The cognitive significance of mirroring is achieved both through the valuation of mirroring itself as informational (i.e., information without symbolic representation) and by treating it as precognitive scaffolding for symbolic activities. One example is the neural theory of concepts, which combines embodied simulation theory with George Lakoff and Mark Johnson's (2005) embodied mind theory of metaphor. They argue that the brain generates concepts, which they see as the building blocks of human reason, by processing information from the sensorimotor system, including from mirror neurons. Rather than the mental, top-down manipulation of symbols, reason is built from the bottom up through the combination of various neural (rather than symbolic) representations— including the neural representations of others' actions derived from mirror neurons. Gallese and art historian David Freedberg (2007) have also advanced a mirror theory of aesthetic experience. In their view embodied simulation allows us to experience images in terms of their emotional and intentional content. While the authors do not discount the influence of cultural context in our aes-

thetic experiences, they depict neurobiology as the basic, fundamental, and universal architect of aesthetic experience that is indifferent to the mediating effects of culture. In Gallese and Freedberg's view, "historical and cultural or contextual factors do not contradict the importance of considering the neural processes that arise in the empathetic understanding of visual works of art" (2007, 202). But perhaps this strategy allows neurobiological research to proceed in isolation from social and cultural analysis, even while it claims authority on cultural matters. It also affirms a biological version of what Pierre Bourdieu calls a "pure gaze," a naive perception that is uninflected by the social position of the perceiver. By contrast, Bourdieu insists that perception of the aesthetic is generated in a social field: "This historical culture functions as a principle of pertinence which enables one to identify, among the elements offered to the gaze, all the distinctive features and only these, by referring them, consciously or unconsciously, to the universe of possible alternatives" (1984, 4).

8 For the response, see Gallese and Sinigaglia 2014.

9 Seeing itself is a social phenomenon. Attention to the physiology of visual perception contests its independence from sociality. Anne Jaap Jacobson (2012) describes how vision involves the binding of elements from early vision, such as color and shape, as well as amodal completion, or the consolidation of "short takes," to produce selective versions of visual imagery. Vision also requires the transition from saccades to objects, which is potentially modulated by attention, which can vary according to interest or expectation, positive and negative valences, and action-relevant factors. Further, visual schemas are elaborated by a learned ability to classify objects and grasp the predictability of their persistence as kinds, conceptualizations based on learned information and social data, and activities of adding-in or filling gaps in what we see; these may depend on various resources such as proprioception, memory, cultural conventions, and other sensory data. Jacobson argues that the need to bind, complete, and bridge gaps in visual input shows how the "community is involved in an individual's knowledge all the way down" (219).

10 Other studies enact mirroring differentially based on the subject position of the observer (such as economic or gender status), learned attributes of the observed (such as their likability), and the appraisal of a situation (Hein and Singer 2008; Singer et al. 2006; Singer and Lamm 2009).

11 The situatedness of mirroring—its material-discursive specificity—ultimately strains its equivalence with simulation. Other models argue that instead of independently achieving simulation, mirror neurons are participants in broader, multimodal processes such as enactive perception (Gallagher 2007; Gallagher and Zahavi 2008; Slaby 2013). Enactive perception, as Gallagher describes it, is an automatic bodily register, but it is skillful; that is, it is developed through the experience of directly engaging and interacting with others. Using an interactionist framework, Jan Slaby argues that mirroring alone does not amount to intersubjective understanding; instead, mirroring is part of a broader event of

mutual interactive engagement, and potentially achieves "a kind of *joint* agency, and *joint* active world-orientation" (2013, 17). For Andrew Murphie, mirror neurons "should not be seen, first, as operators of resemblance—a recognition of a picture—between pregiven agents" (2010, 284). He argues rather that mirror neurons have to be seen in terms of the contact of two different bodies/brains that generates an emergent intersubjective reality.

12 Slaby describes the need for "a stance of *acknowledging*, of *recognizing* the other, both in her (partial) agentive autonomy and in her exposedness as a vulnerable, needy being (Butler 2001). With this, we come to *let her be* in what ultimately remains an inevitable alterity" (2013, 17–18).

Chapter 4. Neurobiology and the Queerness of Kinship

1 In females oxytocin is also made in the corpus luteum, a temporary structure of the ovaries that develops during the luteal phase of the menstrual cycle.

2 A current review of the research (Borrow and Cameron 2012) suggests it is still unclear whether oxytocin is a byproduct of sexual arousal or facilitates it. If the latter, it is unclear whether oxytocin does this in the central or peripheral nervous system, through smooth muscle contraction in male and female reproductive organs.

3 For example, in a seminal study published in *Nature*, Kosfeld et al. (2005) measured the effects of administered oxytocin on men's participation in a trust game involving investments. They reported that those given oxytocin showed more trust of another player than those who were not given oxytocin. In neuroeconomics the research using this methodology is overwhelmingly focused on males because researchers have viewed the menstrual cycle as an unwelcome complicating factor.

4 In the female montane vole, they also found that oxytocin receptor distribution changed twenty-four hours after parturition.

5 It is worth noting here that animal researchers have downplayed the normal variation of parenting styles among mice and other lab animals, and that alloparenting and even cross-species relations of care have been regularly observed in animals whose neurohormonal processes have not been manipulated.

6 On the strong version of this view, articulated, for example, by Larry Young, oxytocin and other neural systems act as biological determinants, directly shaping social life through their role in reproduction. Other accounts, such as Churchland's, are keen to acknowledge the vast cultural diversity in social arrangements and values as well as the plasticity of the oxytocin system.

7 In the two studies described by Rosenblatt and Terkel, both the rats treated with pregnancy hormones and those who were not eventually adopted young that were not their own. The rats administered hormones immediately retrieved pups who moved away from them. In the other study similar behavior developed without the exogenous administration of hormones. Instead, cohabitation and

gradual social interaction in close confines over the course of a week had effects similar to those of hormonal doping. Although Rosenblatt called this a "non-hormonal" mechanism of behavior modification, Feldman's studies of the plasticity of oxytocin point to another interpretation. In *both* of these experiments, adoption could have involved physiological transformations. Rosenblatt was interested in the transformative effects of hormones on relational behavior, but his experiment also pointed a way to thinking about the transformative effects of relations on hormones and other physiological aspects of embodiment.

8 Schiebinger writes that in the eighteenth century, when Carolus Linnaeus classified mammals and named them for milk-producing mammae present in females, breastfeeding was so unpopular that not only aristocrats but also artisans, merchants, and farm families regularly sent their infants away to be nursed. Because wet nurses had inferior nutrition and living conditions and were often feeding multiple babies, the mortality rate was high. Even so, mothers were so reticent to breastfeed, and officials so emphatic that they do, that breastfeeding was the subject of public health campaigns even in the eighteenth century. In Prussia a law was enacted in 1794 to force them. In Jamaica the courts urged European women to halt the practice of cross-racial wet-nursing on grounds of moral depravity. In France anti–wet nursing campaigns extolled the virtues of mother's milk and chastised women for abandoning their natural roles. Leading Enlightenment scholars demanded that mothers must follow their instincts and adhere to the laws of nature. Linnaeus rejected women's explanations for refusing to nurse. He argued, as Schiebinger puts it, "women only pretended to be unable to breastfeed and ridiculed their many 'excuses': that they did not have enough milk, or could not be deprived of fluids precious to their own health, or were overloaded with domestic affairs" (1993, 70). The moralistic claims of the Enlighteners linked breastfeeding to attachment and maternal love, and these to the broader social good. The division of labor between the sexes being negotiated in Europe in the eighteenth century was not based on conserving given arrangements, but on institutionalizing an idealized view of maternity. The maternalized brain model, then, is not the first treatment of the maternal body as the source of all family bonds, even while actual practices suggest otherwise.

9 Andrew and Harvery (2001) point out that breastfeeding rates in the United States and United Kingdom were the highest recorded to date in 2001. In the United States, however, while 69.5 percent of mothers initiated breastfeeding, only 32.5 percent breastfed until six months. In the United Kingdom 78 percent of mothers initiated breastfeeding in 2001. However, the rate of exclusive breastfeeding until six months (as recommended by the World Health Organization) was less than 1 percent.

10 In their study of 160 first-time fathers and mothers (80 couples), the researchers claim to find comparable baseline concentrations of plasma oxytocin in fathers and mothers, activity-dependent plasticity in both, and biological synchrony between coparents as well as between parents and offspring (Feldman 2012).

11 For example, DeVries et al. (1997) claimed that in their lab prairie voles developed partner preferences through cohabitation, regardless of the sex of each vole. In one study the researchers placed same-sex voles in close proximity for twenty-four hours and observed that the voles developed attachments to each other similar to attachments observed in opposite-sex couplings. The researchers identified these attachments as "social bonds" rather than pair bonds, since by standard definition pair bonds must be heterosexual. They also reported that the reproductively naïve females, when given a choice, preferred other females to males. This preference disappears only among females who have reproduced when a male stranger is introduced. To make sense of these findings, DeVries et al. hypothesized that same-sex "social bonds" are important for group formation. This research opens up the recognition of relational variation among voles, while also affirming a heteronormative hierarchy in the context of evolutionary theory.

REFERENCES

Ahmed, Sara. 2006. "Orientations: Toward a Queer Phenomenology." GLQ: A Journal of Lesbian and Gay Studies 12, no. 4: 543–74.

Alaimo, Stacy, and Susan Hekman, eds. 2008. Material Feminisms. Bloomington: Indiana University Press.

Alcoff, Linda Martín. 2006. Visible Identities: Race, Gender and the Self. New York: Oxford University Press.

Anderson, Michael, Michael Richardson, and Anthony Chemero. 2012. "Eroding the Boundaries of Cognition: Implications of Embodiment." Topics in Cognitive Science 4: 1–14.

Andrew, Naomi, and Kate Harvery. 2011. "Infant Feeding Choices: Experience, Self-identity and Lifestyle." Maternal and Child Nutrition 7: 48–60.

Anzaldua, Gloria. 2002. "now let us shift . . . the path of conocimiento . . . inner work, public acts." In This Bridge We Call Home: Radical Visions for Transformation, edited by Gloria Anzaldua and AnaLousie Keating, 540–78. New York: Routledge.

Argent, Gala. 2012. "Toward a Privileging of the NonVerbal: Communication, Corporeal Synchrony, and Transcendence in Humans and Horses." In Experiencing Animal Minds: An Anthology of Animal-Human Encounters, edited by Julie Smith and Robert Mitchell, 111–28. New York: Columbia University Press.

Armstrong, Thomas. 2010. Neurodiversity: Discovering the Gifts of Autism, ADHD, Dyslexia, and Other Brain Differences. Cambridge, MA: Da Capo Press.

Asberg, Cecilia, and Lynda Birke. 2010. "Biology Is a Feminist Issue: Interview with Lynda Birke." European Journal of Women's Studies 17, no. 4: 413–23.

Bagemihl, Bruce. 1999. Biological Exuberance. New York: St. Martin's Press.

Bailey, Nathan, and Marlene Zuk. 2009. "Same-Sex Sexual Behavior and Evolution." Trends in Ecology and Evolution 24, no. 8: 439–46.

Bao, Ai-Min, and Dick F. Swaab. 2011. "Sexual Differentiation of the Human Brain: Relation to Gender Identity, Sexual Orientation and Neuropsychiatric Disorders." Frontiers in Neuroendocrinology 32: 214–26.

Barad, Karen. 2007. Meeting the Universe Halfway: Quantum Physics and the Entanglement of Matter and Meaning. Durham, NC: Duke University Press.

Baril, Alexandre. Forthcoming. "Transness as Debility: Rethinking Intersections between Trans and Disabled Bodies/Identities." Feminist Review, no. 111.

Baron-Cohen, Simon. 2009. "Autism: The Empathizing–Systemizing (e-s) Theory." *Annals of the New York Academy of Sciences* 1156: 68–80.

Baron-Cohen, Simon. 2010. "Empathizing, Systemizing, and the Extreme Male Brain Theory of Autism." In *Sex Differences in the Human Brain, Their Underpinnings and Implications (Progress in Brain Research)*, edited by Ivanka Savic-Berglund, 167–76. New York: Elsevier.

Baron-Cohen, S., R. Knickmeyer, and M. Belmonte. 2005. "Sex Differences in the Brain: Implications for Explaining Autism." *Science* 310: 819–23.

Bartz, Jennifer A., Jamil Zaki, Niall Bolger, and Kevin N. Ochsner. 2011. "Social Effects of Oxytocin in Humans: Context and Person Matter." *Trends in Cognitive Sciences* 15, no. 7: 301–9.

Bates, Elizabeth A., and Jeffrey L. Elman. 2002. "Connectionism and Change." In *Brain Development and Cognition*, 2nd ed., edited by Mark Johnson, Yuko Munakata, and Rick O. Gilmore, 420–40. Oxford: Blackwell.

Baynton, Douglas C. 2013. "Disability and the Justification of Inequality in American History." In *The Disability Studies Reader*, 4th ed., edited by Lennard Davis, 17–33. New York: Routledge.

Berlant, Lauren. 2000. *Intimacy*. Chicago: University of Chicago Press.

Berlucci, Giovanni, and Henry A. Butchel. 2009. "Neuronal Plasticity: Historical Roots and Evolution of Meaning." *Experimental Brain Research* 192, no. 3: 307–19.

Bernhardt, B. C., and T. Singer. 2012. "The Neural Basis of Empathy." *Annual Review of Neuroscience* 35: 1–23.

Birke, Lynda. 1999. *Feminism and the Biological Body*. Edinburgh: Edinburgh University Press.

Blackman, Lisa. 2008. *The Body: The Key Concepts*. New York: Berg.

Blackman, Lisa. 2012. *Immaterial Bodies: Affect, Embodiment, Mediation*. London: Sage.

Blakemore, S. J., D. Bristow, G. Bird, C. Frith, and J. Ward. 2005. "Somatosensory Activations during the Observation of Touch and a Case of Vision–Touch Synaesthesia." *Brain* 128: 1571–83.

Blakeslee, Sandra. 2006. "Cells That Read Minds." *New York Times*, January 10. Accessed August 1, 2012. http://www.nytimes.com/2006/01/10/science/10mirr.html?pagewanted=all&_r=0.

Bliss, Tim, and Terje Lømo. 1973. "Long-Lasting Potentiation of Synaptic Transmission in the Dentate Area of the Anaesthetized Rabbit Following Stimulation of the Perforant Path." *Journal of Physiology* 232: 357–74.

Borrow, Amanda P., and Nicole M. Cameron. 2012. "The Role of Oxytocin in Mating and Pregnancy." *Hormones and Behavior* 61: 266–76.

Bosch, O. J., H. P. Nair, I. D. Neumann, and L. J. Young. 2005. "Depressive-like Behavior Following Isolation from a Female Partner Is Associated with Altered Brain CRF mRNA and HPA Axis Activity in the Male Prairie Vole," Program No. 420.4. Washington, DC: Society for Neuroscience Abstracts.

Bost, Suzanne. 2008. "From Race/Sex/Etc to Glucose, Feeding Tube, and Mourning: The Shifting Matter of Chicana Feminism." In *Material Feminisms*, edited by Stacy Alaimo and Susan Hekman, 340–73. Bloomington: Indiana University Press.

Bourdieu, Pierre. 1984. *Distinction: A Social Critique of the Judgement of Taste*. London: Routledge.

Bourdieu, Pierre. 1990. *The Logic of Practice*. Stanford, CA: Stanford University Press.

Breslaw, Anna. 2013. "LOL: Fox News Accidentally Uses Photo of Gay Couple to Illustrate Traditional Gender Roles." Jezebel.com, February 9. Accessed March 1, 2013. http://jezebel.com/5983020/lol-fox-news-accidentally-uses-photo-of -gay-couple-to-illustrate-traditional-gender-roles.

Brincker, Maria. 2011. "Moving Beyond Mirroring." PhD diss., City University of New York Graduate Center.

Burns, Elaine, Virginia Schmied, Athena Sheehan, and Jennifer Fenwick. 2010. "A Meta-Ethnographic Synthesis of Women's Experience of Breastfeeding." *Maternal and Child Nutrition* 6, no. 3 (July): 201–19.

Butler, Judith. 1990. "Performative Acts and Gender Constitution: An Essay in Phenomenology and Feminist Theory." In *Performing Feminisms: Feminist Critical Theory and Theatre*, edited by Sue-Ellen Case, 270–82. Baltimore: Johns Hopkins University Press.

Butler, Judith. 1993. *Bodies That Matter: On the Discursive Limits of "Sex."* New York: Routledge.

Butler, Judith. 2004. *Undoing Gender*. New York: Routledge.

Butler, Judith. 2010. *Frames of War: When Is Life Grievable?* London: Verso.

Campbell, Neil. 2009. Interview with Patricia Churchland. *Campbell Biology 4th Edition*, edited by Eric J. Simon, Jane B. Reece, and Jean L. Dickey. San Francisco: Benjamin Cummings.

Caramazza, A., S. Anzellotti, L. Strand, and A. Lingnau. 2014. "Embodied Cognition and Mirror Neurons: A Critical Assessment." *Annual Review of Neuroscience* 37: 1–15.

Casey, B. J., Rebecca M. Jones, and Todd A. Hare. 2008. "The Adolescent Brain." *Annals of the New York Academy of Sciences* 1124: 111–26.

Catmur, Caroline, Vincent Walsh, and Cecilia Heyes. 2007. "Sensorimotor Learning Configures the Mirror Neuron System." *Current Biology* 17, no. 17: 1527–31.

Cheah, Pheng. 1996. "Mattering." *Diacritics* 26, no. 1: 108–39.

Churchland, Patricia Smith. 1986. *Neurophilosophy: Toward a Unified Science of the Mind-Brain*. Cambridge, MA: MIT Press.

Churchland, Patricia. 2011. *Braintrust: What Neuroscience Tells Us about Morality*. Princeton, NJ: Princeton University Press.

Churchland, Patricia S., and Piotr Winkielman. 2012. "Modulating Social Behavior with Oxytocin: How Does It Work? What Does It Mean?" *Hormones and Behavior* 61, no. 3: 392–99.

Clark, Andy. 1998. "Embodiment and the Philosophy of Mind." In *Current Issues in Philosophy of Mind: Royal Institute of Philosophy Supplement 43*, edited by Anthony O'Hear, 35–52. New York: Cambridge University Press.

Clark, Andy. 2001. *Mindware: An Introduction to the Philosophy of Cognitive Science.* New York: Oxford University Press.

Clark, Andy. 2004a. "Author's Response to 'We Have Always Been . . . Cyborgs.'" *Metascience* 13: 169–81.

Clark, Andy. 2004b. *Natural-Born Cyborgs: Minds, Technologies, and the Future of Human Intelligence.* Oxford: Oxford University Press.

Clark, Andy. 2007. "Re-Inventing Ourselves: The Plasticity of Embodiment, Sensing, and Mind." *Journal of Medicine and Philosophy* 32, no. 3: 263–82.

Clark, Andy. 2008a. "Pressing the Flesh: A Tension in the Study of the Embodied, Embedded Mind?" *Philosophy and Phenomenological Research* 76, no. 1: 37–59.

Clark, Andy. 2008b. *Supersizing the Mind: Embodiment, Action, and Cognitive Extension.* Oxford: Oxford University Press.

Clifford, Erin. 1999. "Neural Plasticity: Merzenich, Taub, Greenough." *Harvard Brain* 6, no. 1: 16–20.

Clough, Patricia T. 2007. Introduction. In *The Affective Turn: Theorizing the Social*, edited by Patricia Clough with Jean Halley, 1–33. Durham, NC: Duke University Press.

Clough, Patricia T. 2010. "The Affective Turn: Political Economy, Biomedia, and Bodies." In *The Affect Theory Reader*, edited by Melissa Gregg and Gregory Seigworth, 206–28. Durham, NC: Duke University Press.

Clough, Patricia T. 2014. "Sociality, Non-Conscious Processes and the Neurotypical Subject." *Contemporary Sociology: A Journal of Reviews* 43, no. 5 (September): 645–49.

Clough, Patricia, Karen Gregory, Benjamin Haber, and R. Joshua Scannell. 2015. "The Datalogical Turn." In *Nonrepresentational Methodologies: Re-envisioning Research*, edited by Phillip Vannini, 146–64. London: Routledge.

Colebrook, Claire. 2000. "Questioning Representation." *SubStance* 29, no 2: 47–67.

Colebrook, Claire. 2008. "On Not Becoming Man: The Materialist Politics of Un-actualized Potential." In *Material Feminisms*, edited by Stacy Alaimo and Susan Hekman, 52–84. Bloomington: Indiana University Press.

Collins, Patricia Hill. 1990. *Black Feminist Thought: Knowledge, Consciousness, and the Politics of Empowerment.* New York: HarperCollins.

Collins, Patricia Hill. 2000. *Black Feminist Thought: Knowledge, Consciousness, and the Politics of Empowerment.* 2nd ed. New York: Routledge.

Connolly, William. 2002. *Neuropolitics: Thinking, Culture, Speed.* Minneapolis: University of Minnesota Press.

Connolly, William. 2010. *A World of Becoming.* Durham, NC: Duke University Press.

Connolly, William. 2011. "The Complexity of Intention." *Critical Inquiry* 37, no. 4: 791–98.

Cook, R., G. Bird, C. Catmur, C. Press, and C. Heyes. 2014. "Mirror Neurons: From Origin to Function." *Behavioral and Brain Sciences* 37, no. 2: 177–92.

Coole, Diana, and Samantha Frost. 2010. "Introducing the New Materialisms." In *New Materialisms: Ontology, Agency and Politics*, edited by Samantha Frost and Diana Coole, 1–46. Durham, NC: Duke University Press.

Coplan, Amy, and Peter Goldie, eds. 2014. *Empathy: Philosophical and Psychological Perspectives*, 45–57. Oxford: Oxford University Press.

Crenshaw, Kimberlé. 1989. "Demarginalizing the Intersection of Race and Sex: A Black Feminist Critique of Antidiscrimination Doctrine, Feminist Theory and Antiracist Politics." *University of Chicago Legal Forum* 140: 139–67.

Crenshaw, Kimberlé. 1991. "Mapping the Margins: Intersectionality, Identity Politics, and Violence against Women of Color." *Stanford Law Review* 43, no. 6: 1241–99.

Crick, Francis. 1994. *The Astonishing Hypothesis: The Scientific Search for the Soul*. New York: Touchstone.

Cunningham, Kim. 2012. "Should We Be Triggered? NeuroGovernance in the Future/(Tense)." *Social Text/Periscope*. Accessed August 1, 2014. http://www.socialtextjournal.org/periscope/neuroculture.

Damasio, Antonio. 1994. *Decartes' Error: Emotions, Reason and the Human Brain*. New York: Harper Perennial.

Damasio, Antonio. 1996. "The Somatic Marker Hypothesis and the Possible Functions of the Prefrontal Cortex." *Philosophical Transactions of the Royal Society* 351: 1413–20.

Damasio, Antonio, and Kasper Meyer. 2008. "Behind the Looking Glass." *Nature* 454: 167–68.

D'Angiulli Amedeo, Anthony Herdman, David Stapells, Clyde Hertzman. 2008. "Children's Event-Related Potentials of Auditory Selective Attention Vary with Their Socioeconomic Status." *Neuropsychology* 22, no. 3: 293–300.

Davis, Lennard. *Enforcing Normalcy: Disability, Deafness and the Body*. London: Verso, 1995.

Davis, Noela. 2009. "New Materialism and Feminism's Anti-Biologism: A Response to Sara Ahmed." *European Journal of Women's Studies* 16, no. 1: 67–80.

Davis, Noela. 2014. "Politics Materialized: Rethinking the Materiality of Feminist Political Action through Epigenetics." *Women: A Cultural Review* 25, no. 1: 62–77.

Debes, Remy. 2010. "Which Empathy? Limitations in the Mirrored 'Understanding' of Emotion." *Synthese* 175: 219–39.

DeFelipe, Javier. 2002. "Sesquicentenary of the Birthday of Santiago Ramon y Cajal, the Father of Modern Neuroscience." *Trends in Neurosciences* 25, no. 9: 481–84.

DeFelipe, Javier. 2006. "Brain Plasticity and Mental Processes: Cajal Again." *Nature Reviews Neuroscience* 7: 811–17.

Deleuze, Gilles, and Felix Guattari. 1987. *A Thousand Plateaus*, translated by Brian Massumi. Minneapolis: University of Minnesota Press.

den Hertog, C., A. de Groot, and P. van Dongen. 2001. "History and Use of Oxytocics." *European Journal of Obstetrics, Gynecology, and Reproductive Biology* 94, no. 1: 8–12.

DeVries, A. Courtney, Camron L. Johnson, C. Sue Carter. 1997. "Familiarity and Gender Influence Social Preferences in Prairie Voles." *Canadian Journal of Zoology* 23: 107–17.

Dewsbury, J-D. 2011. "The Deleuze-Guattarian Assemblage: Plastic Habits." *Area* 43, no. 2: 148–53.

Diamond, Adele, W. Steven Barnett, Jessica Thomas, and Sarah Munro. 2007. "Preschool Program Improves Cognitive Control." *Science* 318, no. 5855 (November): 1387–88.

Diamond, Adele, and Kathleen Lee. 2011. "Interventions Shown to Aid Executive Function Development in Children 4–12 Years Old." *Science* 333, no. 6045: 959–64.

Dolphijn, Rick, and Iris van der Tuin. 2012. "Matter Feels, Converses, Suffers, Desires, Yearns and Remembers: Interview with Karen Barad." *New Materialism: Interviews and Cartographies*. Ann Arbor, MI: Open Humanities Press. Accessed March 10, 2015. http://dx.doi.org/10.3998/ohp.11515701.0001.001.

Dumit, Joe. 2004. *Picturing Personhood: Biomedical Scans and Personal Identity*. Princeton, NJ: Princeton University Press.

Dussauge, Isabelle, and Anelis Kaiser. 2012a. "Re-Queering the Brain." In *Neurofeminism: Issues at the Intersection of Feminist Theory and Cognitive Science*, edited by Robyn Bluhm, Anne Jaap Jacobsen, and Heidi Lene Maiborn, 121–44. New York: Palgrave Macmillan.

Dussauge, Isabelle, and Anelis Kaiser. 2012b. "Neuroscience and Sex/Gender." *Neuroethics* 5, no. 3: 211–15.

Einstein, Gillian. 2012. "Situated Neuroscience: Exploring a Biology of Diversity." In *Neurofeminism: Issues at the Intersections of Feminist Theory and Cognitive Science*, edited by Robyn Bluhm, Anne Jaap Jacobsen, and Heidi Lene Maiborn, 145–74. New York: Palgrave Macmillan.

Eliot, Lise. 2009. "Girl Brain, Boy Brain?: The Two Are Not the Same, but New Work Shows Just How Wrong It Is to Assume That All Gender Differences Are 'Hardwired.'" *Scientific American*, September 8. Accessed August 1, 2013. http://www.scientificamerican.com/article/girl-brain-boy-brain/.

Eng, David. 2010. *The Feeling of Kinship: Queer Liberalism and the Racialization of Intimacy*. Durham, NC: Duke University Press.

Enticott, P. G., H. A. Kennedy, N. J. Rinehart, J. L. Bradshaw, B. J. Tonge, Z. J. Daskalakis, and P. B. Fitzgerald. 2013. "Interpersonal Motor Resonance in Autism Spectrum Disorder: Evidence against a Global 'Mirror System' Deficit." *Frontiers in Human Neuroscience* 7, no. 218: 1–8.

Farah, Martha J., Kimberly G. Noble, and Hallam Hurt. 2006. "Poverty, Privilege, and Brain Development: Empirical Findings and Ethical Implications." In *Neuroethics: Defining the Issues in Theory, Practice and Policy*, edited by Judy Illes, 277–89. Oxford: Oxford University Press.

Farah, Martha J., David M. Shera, Jessica H. Savage, Laura Betancourt, Joan M. Giannettac, Nancy L. Brodsky, Elsa K. Malmud, and Hallam Hurt. 2006. "Childhood Poverty: Specific Associations with Neurocognitive Development." *Brain Research* 1110: 166–74.

Fausto-Sterling, Anne. 1992. "Building Two-Way Streets: The Case of Feminism and Science." *NWSA Journal* 4, no. 3: 336–49.

Fausto-Sterling, Anne. 2012. *Sex/Gender: Biology in a Social World*. New York: Routledge.

Fausto-Sterling, Anne, M. Blackless, A. Charuvastra, A. Derryck, K. Lauzanne, and E. Lee. 2000. "How Sexually Dimorphic Are We? Review and Synthesis." *American Journal of Human Biology* 12: 151–66.

Fecteau, Shirley, Jose Maria Tormos, Massimo Gangitano, Hugo Theoret, and Alvaro Pascual-Leone. 2010. "Modulation of Cortical Motor Outputs by the Symbolic Meaning of Visual Stimuli." *European Journal of Neuroscience* 32: 172–77.

Feldman, Ruth. 2007. "Parent–Infant Synchrony Biological Foundations and Developmental Outcomes." *Current Directions in Psychological Science* 16, no. 6: 340–44.

Feldman, Ruth. 2012. "Oxytocin and Affiliation in Humans." *Hormones and Behavior* 61: 380–91.

Ferguson, Jennifer N., Larry J. Young, Elizabeth F. Hearn, Martin M. Matzuk, Thomas R. Insel, and James T. Winslow. 2000. "Social Amnesia in Mice Lacking the Oxytocin Gene." *Nature Genetics* 25, no. 3: 284–88.

Fine, Cordelia. 2010. "From Scanner to Sound Bite: Issues in Interpreting and Reporting Sex Differences in the Brain." *Current Directions in Psychological Science* 19: 280–83.

Fine, Cordelia. 2011. *Delusions of Gender: How Our Minds, Society, and Neuroscience Create Difference*. New York: W. W. Norton.

Fine, Cordelia, Rebecca Jordan-Young, Anelis Kaiser, and Gina Rippon. 2013. "Plasticity, Plasticity, Plasticity . . . and the Rigid Problem of Sex." *Trends in Cognitive Science* 17, no. 11: 550–51.

Fleming, Alison. 2007. "The Three Faces of Jay Rosenblatt." *Developmental Psychobiology* 49, no. 1: 2–11.

Foucault, Michel. 1975. *Discipline and Punish: The Birth of the Prison*. New York: Random House.

Foucault, Michel. 1979. *The History of Sexuality Volume 1: An Introduction*. London: Allen Lane.

Foucault, Michel. 2009. *Security, Territory, Population: Lectures at the Collège de France 1977–78*. Translated by Graham Burchell. New York: Picador.

Franks, Arthur. 2004. *The Renewal of Generosity: Illness, Medicine, and How to Live*. Chicago: University of Chicago Press.

Franks, David. 2010. *Neurosociology: The Nexus between Neuroscience and Social Psychology*. New York: Springer.

Freedberg, David, and Vittorio Gallese. 2007. "Motion, Emotion and Empathy in Esthetic Experience." *Trends in Cognitive Neuroscience* 11, no. 5: 197–203.

Freeman, Elizabeth. 2010. *Time Binds: Queer Temporalities, Queer Histories.* Durham, NC: Duke University Press.

Fritsch, Jane. 2000. "The Diallo Verdict: The Overview; 4 Officers in Diallo Shooting Are Acquitted of All Charges." *New York Times*, February 26.

Fuchs, Thomas. 2005. "Overcoming Dualism." *Philosophy, Psychiatry, and Psychology* 12, no. 2: 115–17.

Fuchs, Thomas. 2009. "Embodied Cognitive Neuroscience and Its Consequences for Psychiatry." *Poiesis Prax* 6, nos. 3–4: 219–33.

Fullagar, Simone. 2009. "Negotiating the Neurochemical Self: Anti-Depressant Consumption in Women's Recovery from Depression." *Health (London)* 13, no. 4: 389–406.

Gallagher, Shaun. 2007. "Simulation Trouble." *Social Neuroscience* 2: 353–65.

Gallagher, Shaun, and Dan Zahavi. 2008. *The Phenomenological Mind: An Introduction to Philosophy of Mind and Cognitive Science.* London: Routledge.

Gallese, Vittorio. 2001. "The 'Shared Manifold' Hypothesis: From Mirror Neurons to Empathy." *Journal of Consciousness Studies* 8, nos. 5–7: 33–50.

Gallese, Vittorio. 2003. "The Roots of Empathy: The Shared Manifold Hypothesis and the Neural Basis of Intersubjectivity." *Psychopathology* 36: 171–80.

Gallese, Vittorio. 2009. "Mirror Neurons, Embodied Simulation, and the Neural Basis of Social Identification." *Psychoanalytic Dialogues* 19: 519–36.

Gallese, Vittorio. 2014. "Bodily Selves in Relation: Embodied Simulation as Second-person Perspective on Intersubjectivity." *Philosophical Transcripts of the Royal Society B: Biological Sciences* 369, no. 1644. doi 10.1098/rstb.2013.0177.

Gallese, Vittorio, and Alvin Goldman. 1998. "Mirror Neurons and the Simulation Theory of Mind-Reading." *Trends in Cognitive Sciences* 2: 493–501.

Gallese, Vittorio, Christian Keysers, and Giacomo Rizzolatti. 2004. "A Unifying View of the Basis of Social Cognition." *Trends in Cognitive Sciences* 8, no. 9: 396–403.

Gallese, Vittorio, and George Lakoff. 2005. "The Brain's Concepts: The Role of the Sensory-Motor System in Conceptual Knowledge." *Cognitive Neuropsychology* 22, nos. 3/4: 455–79.

Gallese, Vittorio, Magali Rochat, Corrado Sinigaglia, and Giuseppe Cossu. 2009. "Motor Cognition and Its Role in the Phylogeny and Ontogeny of Action Understanding." *Developmental Psychology* 45, no. 1: 103–13.

Gallese, Vittorio, M. J. Rochat, and C. Berchio. 2013. "The Mirror Mechanism and Its Potential Role in Autism Spectrum Disorder." *Developmental Medicine and Child Neurology* 55 no. 1: 15–22.

Gallese, Vittorio, and Corrado Sinigaglia. 2014. "Understanding Action with the Motor System." *Behavioral and Brain Sciences* 37 no. 2: 199–200.

Garland-Thomson, Rosemarie. 2001. "Reshaping, Re-thinking, Re-defining: Feminist Disability Studies." Barbara Waxman Fiduccia Papers on Women and Girls with Disabilities Center for Women Policy Studies. Washington, DC: Center for Women Policy Studies. Accessed August 1, 2014. http://www.centerwomen policy.org/pdfs/DIS2.pdf.

Garland-Thomson, Rosemarie. 2002. "The Politics of Staring: Visual Rhetorics of Disability in Popular Photography." In *Disability Studies: Enabling the Humanities*, edited by Sharon L. Snyder, 56–75. New York: Modern Language Association Press.

Garland-Thomson, Rosemarie. 2011. "Misfits: A Feminist Materialist Disability Concept." *Hypatia* 26, no. 3: 591–609.

Geertz, Clifford. 1973. *The Interpretation of Cultures: Selected Essays*. New York: Basic Books.

Gerdes, Karen E., Cynthia A. Lietz, and Elizabeth A. Segal. 2011. "Measuring Empathy in the 21st Century: Development of an Empathy Index Rooted in Social Cognitive Neuroscience and Social Justice." *Social Work Research* 35, no. 2: 83–93.

Gibson, James. 1966. *The Senses Considered as Perceptual Systems*. Boston: Houghton Mifflin.

Giedd, Jay N. 2004. "Structural Magnetic Resonance Imaging of the Adolescent Brain." *Annals of the New York Academy of Sciences* 1021 (June): 77–85.

Gil, M., R. Bhatt, K. B. Picotte, and E. M. Hull. 2013. "Sexual Experience Increases Oxytocin Receptor Gene Expression and Protein in the Medial Preoptic Area of the Male Rat." *Psychoneuroendocrinology* 38, no. 9: 1688–97.

Gillis-Buck, Eva M., and Sarah S. Richardson. 2014. "Autism as a Biomedical Platform for Sex Differences Research." *BioSocieties* 9, no. 3: 262–83.

Gladwell, Malcolm. 2005. *Blink: The Power of Thinking without Thinking*. New York: Back Bay Books.

Glannon, Walter. 2002. "Depression as a Mind-Body Problem." *Philosophy, Psychiatry and Psychology* 9, no. 3: 243–54.

Glannon, Walter. 2009. "Our Brains Are Not Us." *Bioethics* 23, no. 6: 321–29.

Goddings, Anne-Lise, L. Menzies, I. Dumontheil, E. Garrett, S. J. Blakemore, and R. M. Viner. 2015. "The Relationship between Pubertal Status and Neural Activity during Risky Decision-making in Male Adolescents Using fMRI." *Archives of Disease in Childhood* 100, Supplement 3: A64.

Goldman, Alvin. 2009. "Mirroring, Simulating and Mindreading." *Mind and Language* 24, no. 2: 235–52.

Goldman, Alvin, and Karen Shanton. Forthcoming. "The Case for Simulation Theory." In *Handbook of "Theory of Mind,"* edited by A. Leslie and T. German. New York: Psychology Press. Accessed July 13, 2015. http://www.nyu.edu/gsas/dept/philo/faculty/block/M&L2010/Papers/Goldman.pdf.

Gordon, Robert M. 2009. "Folk Psychology as Mental Simulation." In *The Stanford Encyclopedia of Philosophy*, edited by Edward N. Zalta. December 8, 1997. Accessed May 6, 2013. http://plato.stanford.edu/entries/folkpsych-simulation.

Grafton, Scott T. 2009. "Embodied Cognition and the Simulation of Action to Understand Others." *Annals of the New York Academy of Sciences* 1156: 97–117.

Graustella, Adam J., and Colin MacLeod. 2012. "A Critical Review of the Influence of Oxytocin Nasal Spray on Social Cognition in Humans: Evidence and Future Directions." *Hormones and Behavior* 16, no. 3: 410–18.

Green, Adam. 2009. "Mirror Neurons, Simulation, and Goldman." *History and Philosophy of Psychology* 11, no. 2: 1–11.

Grosz, Elizabeth. 1994. *Volatile Bodies: Towards a Corporeal Feminism*. Bloomington: Indiana University Press.

Grosz, Elizabeth. 2004. *The Nick of Time: Politics, Evolution, and the Untimely*. Durham, NC: Duke University Press.

Grosz, Elizabeth. 2005. *Time Travels: Feminism, Nature, Power*. Durham, NC: Duke University Press.

Gumy, Christel. 2014. "The Gendered Tools of the Construction of a Unisex 'Adolescent Brain.'" In *Gendered Neurocultures: Feminist and Queer Perspectives on Current Brain Discourses*, edited by Sigrid Schmitz and Grit Höppner, 257–72. Vienna: Zaglossus.

Hackman, Daniel A., Laura M. Betancourt, Nancy L. Brodsky, Hallum Hurt, Martha J. Farah. 2012. "Neighborhood Disadvantage and Adolescent Stress Reactivity." *Frontiers in Human Neuroscience* 6, article 277: 1–11.

Hackman, Daniel A., and Martha J. Farah. 2009. "Socioeconomic Status and the Developing Brain." *Trends in Cognitive Sciences* 13, no. 2: 65–73.

Hackman, Daniel A., Martha J. Farah, and Michael J. Meaney. 2010. "Socioeconomic Status and the Brain: Mechanistic Insights from Human and Animal Research." *Nature Reviews Neuroscience* 11, no. 9: 651–59.

Halberstam, Judith. 2005. *In a Queer Time and Place: Transgender Bodies, Subcultural Lives*. New York: New York University Press.

Hamilton, Antonia F., R. M. Brindley, and U. Frith. 2007. "Imitation and Action Understanding in Autistic Spectrum Disorders: How Valid Is the Hypothesis of a Deficit in the Mirror Neuron System?" *Neuropsychologia* 45: 1859–68.

Haraway, Donna. 1988. "Situated Knowledges: The Science Question in Feminism and the Privilege of Partial Perspective." *Feminist Studies* 14, no. 3 (autumn): 575–99.

Haraway, Donna. 1991. *Simians, Cyborgs, and Women: The Reinvention of Nature*. New York: Routledge.

Haraway, Donna. 1997. *Modest_Witness@Second_Millennium.FemaleMan_Meets _OncoMouse*. New York: Routledge.

Harding, Sandra. 1986. *The Science Question in Feminism*. Ithaca, NY: Cornell University Press.

Hartsock, Nancy. 1983. "The Feminist Standpoint: Developing a Specifically Feminist Historical Materialism." In *Discovering Reality: Feminist Perspectives on Epistemology and Methodology*, edited by Sandra Harding and Merrill Hintikka, 283–310. New York: Kluwer.

Hauptmann, Deborah. 2010. "Introduction: Architecture and Mind in the Age of Communication and Information." In *Cognitive Architecture: From Biopolitics to Noopolitics,* edited by Deborah Hauptmann and Warren Neidich, 11–44. Rotterdam: Delft School of Design.

Hayles, N. Katherine. 1993. "Virtual Bodies and Flickering Signifiers." *October* 66 (autumn): 69–91.

Hein, Grit, and Tania Singer. 2008. "I Feel How You Feel But Not Always: The Empathic Brain and Its Modulation." *Current Opinion in Neurobiology* 18: 153–58.

Heinrichs, Markus, Bernadette von Dawans, and Gregor Domes. 2009. "Oxytocin, Vasopressin, and Human Social Behavior." *Frontiers in Neuroendocrinology* 30: 548–57.

Hemmings, Clare. 2005. "Invoking Affect: Cultural Theory and the Ontological Turn." *Cultural Studies* 19, no. 5: 548–67.

Hemmings, Clare. 2012. "Affective Solidarity: Feminist Reflexivity and Political Transformation." *Feminist Theory* 13, no. 2: 147–61.

Herbert, J. 1994. "Oxytocin and Sexual Behavior." *British Medical Journal* 309: 891–92.

Heyes, Cecilia. 2010a. "Mesmerizing Mirror Neurons." *NeuroImage* 51: 789–91.

Heyes, Cecilia. 2010b. "Where Do Mirror Neurons Come From?" *Neuroscience and Biobehavioral Reviews* 34: 575–83.

Hickok, Gregory. 2009. "Eight Problems for the Mirror Neuron Theory of Action Understanding in Monkeys and Humans." *Journal of Cognitive Neuroscience* 21, no. 7: 1229–43.

Hickok, Gregory. 2014. *The Myth of Mirror Neurons: The Real Neuroscience of Communication and Cognition.* New York: W. W. Norton.

Hird, Myra J. 2004. "Chimerism, Mosaicism and the Cultural Construction of Kinship." *Sexualities* 7, no. 2: 217–32.

Hird, Myra J. 2009. "Feminist Engagements with Matter." *Feminist Studies* 35, no. 2: 329–46.

Ho, Christina, and Ingrid Schraner. 2004. "Feminist Standpoints, Knowledge and Truth: A Literature Review." School of Economics and Finance Working Paper Series. Sydney: University of Western Sydney.

Hubel, David H., and Torsten N. Weisel. 1970. "The Period of Susceptibility to the Physiological Effects of Unilateral Eye Closure in Kittens." *Journal of Physiology* 206, no. 2: 419–36.

Hubel, David H., and Torsten N. Weisel. 1998. "Early Exploration of the Visual Cortex." *Neuron* 20, no. 3 (March): 401–12.

Hurley, Kristen M., Maureen M. Black, Mia A. Papas, and Anna M. Quigg. 2008. "Variation in Breastfeeding Behaviours, Perceptions, and Experiences by Race/ Ethnicity among a Low-Income Statewide Sample of Special Supplemental Nutrition Program for Women, Infants, and Children (WIC) Participants in the United States." *Maternal and Child Nutrition* 4: 95–105.

Hyde, Janet S. 2014. "Gender Similarities and Differences." *Annual Review Psychology* 65: 373–98.

Iacoboni, Marco. 2008. *Mirroring People: The Science of How We Connect with Others.* New York: Farrar, Straus, and Giroux.

Iacoboni, Marco. 2009. "Imitation, Empathy, and Mirror Neurons." *Annual Review of Psychology* 60: 653–70.

Iacoboni, Marco. 2011. "Within Each Other: Neural Mechanisms for Empathy in the Primate Brain." In *Empathy: Philosophical and Psychological Perspectives*, edited by Amy Coplan and Peter Goldie, 45–57. Oxford: Oxford University Press.

Iacoboni, Marco, I. Molnar-Szakacs, V. Gallese, G. Buccino, J. C. Mazziotta, and G. Rizzolatti. 2005. "Grasping the Intentions of Others with One's Own Mirror Neuron System." *PLoS Biology* 3, no. 3: e79.

Insel, Thomas R., Larry Young, and Zuoxin Wang. 1997. "Central Oxytocin and Reproductive Behaviours." *Reviews of Reproduction* 2: 28–37.

Insel, Tom R., and Lawrence E. Shapiro. 1992. "Oxytocin Receptor Distribution Reflects Social Organization in Monogamous and Polygamous Voles." *Proceedings of the National Academy of Sciences of the United States of America* 89, no. 13: 5981–85.

Iwakuma, Miho. 2006. "The Body as Embodiment: An Investigation of the Body by Merleau-Ponty." In *Disability/Postmodernity: Embodying Disability Theory*, edited by Mairian Corker and Tom Shakespeare, 76–87. London: Continuum.

Jacob, Pierre. 2009. "A Philosopher's Reflections on the Discovery of Mirror Neurons." *Topics in Cognitive Science* 1: 570–95.

Jacobson, Anne Jaap. 2012. "Seeing as Social Phenomenon: Feminist Theory and the Cognitive Sciences." In *Neurofeminism: Issues at the Intersection of Feminism and Cognitive Science*, edited by Robyn Bluhm, Anne Jaap Jacobson, and Heidi Lene Maiborn, 216–29. New York: Palgrave Macmillan.

James, William. [1890] 1950. *The Principles of Psychology Volume 1*. New York: Henry Holt.

Joel, Daphna. 2014. "Sex, Gender and Brain: A Problem of Conceptualization." In *Gendered Neurocultures: Feminist and Queer Perspectives on Current Brain Discourses*, edited by S. Schmitz and G. Höppner, 169–86. Vienna: Zaglossus.

Johnson, Mark, and Tim Rohrer. 2007. "We Are Live Creatures: Embodiment, American Pragmatism, and the Cognitive Organism." In *Body, Language, and Mind, Vol. 1: Embodiment*, edited by Tom Ziemke, Jordan Zlatev, and Roslyn M. Frank, 17–54. Berlin: Mouton de Gruyter.

Johnson, Sara B., and Jay N. Giedd. 2014. "Normal Brain Development and Child/Adolescent Policy." In *Handbook of Neuroethics*, edited by Jens Clausen and Neil Levy, 1721–35. New York: Springer.

Jordan-Young, Rebecca. 2010. *Brainstorm: Flaws in the Science of Sex Difference*. Cambridge, MA: Harvard University Press.

Jordan-Young, Rebecca. 2014. "Fragments for the Future: Tensions and New Directions," from "Neurocultures—NeuroGenderings II." In *Gendered Neurocultures: Feminist and Queer Perspectives on Current Brain Discourses*, edited by Sigrid Schmitz and Grit Höppner, 373–95. Vienna: Zaglossus.

Jordan-Young, Rebecca, and Raffaella I. Rumiati. 2012. "Hardwired for Sexism? Approaches to Sex/Gender in Neuroscience." *Neuroethics* 5, no. 3: 305–15.

Joyner, Chuck. 2009. "Fighting the Contagious Fire Phenomenon." January 29. Accessed September 1, 2013. http://www.survivalsciences.com/.

Kaiser, Anelis. 2016. "Gender Matters and Gender Materialities in the Brain." In *Mattering: Feminism, Science and Materialism*, edited by Victoria Pitts-Taylor. New York: New York University Press.

Keen, Suzanne. 2006. "A Theory of Narrative Empathy." *Narrative* 14, no. 3: 207–36.

Kelleher, Christa M. 2006. "The Physical Challenges of Early Breastfeeding." *Social Science and Medicine* 63, no. 10: 2727–38.

Kemp, Andrew, and Adam Guastella. 2011. "The Role of Oxytocin in Human Affect: A Novel Hypothesis." *Current Directions in Psychological Science* 20, no. 4: 222–31.

Keysers, Christian, Jon H. Kaas, and Valeria Gazzola. 2010. "Somatosensation in Social Perception." *Nature Reviews Neuroscience* 11: 417–28.

Kilner, J. M., and R. N. Lemon. 2013. "What We Currently Know about Mirror Neurons." *Current Biology* 23: R1057–62.

Kirby, Vicki. 2008. "Subject to Natural Law: A Meditation on the 'Two Cultures' Problem." *Australian Feminist Studies* 23, no. 55: 5–17.

Kirby, Vicki. 2011. *Quantum Anthropologies*. Durham, NC: Duke University Press.

Kishiyama, Mark M., W. Thomas Boyce, Amy M. Jimenez, Lee M. Perry, and Robert T. Knight. 2008. "Socioeconomic Disparities Affect Prefrontal Function in Children." *Journal of Cognitive Neuroscience* 21, no. 6: 1106–15.

Kosfeld, M., M. Heinrichs, P. J. Zak, U. Fischbacher, and E. Fehr. 2005. "Oxytocin Increases Trust in Humans." *Nature* 435: 673–76.

Kraus, Cynthia. 2012. "Critical Studies of the Sexed Brain: A Critique of What and for Whom?" *Neuroethics* 5, no. 3: 247–59.

Krause, Rainer. 2010. "An Update on Primary Identification, Introjection, and Empathy." *International Forum of Psychoanalysis* 19: 138–43.

Kuhn, Thomas S. 1962. *The Structure of Scientific Revolutions*. Chicago: University of Chicago Press.

Lakoff, George, and Mark Johnson. 1980. *Metaphors We Live By*. Chicago: University of Chicago Press.

Lakoff, George, and Mark Johnson. 1999. *Philosophy in the Flesh: The Embodied Mind and Its Challenge to Western Thought*. New York: Basic Books.

Landecker, Hannah. 2011. "Food as Exposure: Nutritional Epigenetics and the New Metabolism." *BioSocieties* 6: 167–94.

Lane, Riki. 2009. "Trans as Bodily Becoming: Rethinking the Biological as Diversity, Not Dichotomy." *Hypatia* 24, no. 3: 136–57.

Latour, Bruno. 2004. "Why Has Critique Run out of Steam?: From Matters of Fact to Matters of Concern." *Critical Inquiry* 30, no. 2: 225–48.

Ledford, H. 2008. "'Monogamous' Voles in Love-Rat Shock." *Nature* 451, no. 7179: 617.

Leff, Ellen W., Margaret P. Gagne, and Sandra C. Jefferis. 1994. "Maternal Perceptions of Successful Breastfeeding." *Journal of Human Lactation* 10, no. 2: 99–104.

Lehrer, Jonah. 2008. "The Mirror Neuron Revolution: Explaining What Makes Humans Social." *Scientific American*, July 1. Accessed August 1, 2013. http://www.scientificamerican.com/article/the-mirror-neuron-revolut/.

Leighton, Jane, Geoffrey Bird, Tony Charman, and Cecilia Heyes. 2008. "Weak Imitative Performance Is Not Due to a Functional 'Mirroring' Deficit in Adults with Autism Spectrum Disorders." *Neuropsychologia* 46: 1041–49.

Leys, Ruth. 2011. "The Turn to Affect: A Critique." *Critical Inquiry* 37, no. 3: 434–72.

Leys, Ruth. 2012. "'Both of Us Disgusted in My Insula': Mirror Neuron Theory and Emotional Empathy." Nonsite.org. Accessed July 1, 2014.

Lizardo, Omar. 2007. "Mirror Neurons, Collective Objects and the Problem of Transmission: Reconsidering Stephen Turner's Critique of Practice Theory." *Journal for the Theory of Social Behaviour* 37: 319–50.

Lizardo, Omar. 2014. "Beyond the Comtean Schema: The Sociology of Culture and Cognition versus Cognitive Social Science." *Sociological Forum* 29, no. 4: 983–89.

Longino, Helen. 2010. "Feminist Epistemology at Hypatia's 25th Anniversary." *Hypatia* 25, no. 4: 733–41.

Luders, Eileen, Paul M. Thompson, and Arthur W. Toga. 2010. "The Development of the Corpus Callosum in the Healthy Human Brain." *Journal of Neuroscience* 30, no. 33: 10985–90.

MacDonald, K., and T. M. Macdonald. 2010. "The Peptide That Binds: A Systematic Review of Oxytocin and Its Prosocial Effects in Humans." *Harvard Review of Psychiatry* 18, no. 1: 1–21.

MacLean, Heather. 1988. "Women's Experience of Breastfeeding: A Much Needed Perspective." *Health Promotion International* 3, no. 4: 361–70.

Macmillan, Malcolm. 2002. *An Odd Kind of Fame: Stories of Phineas Gage*. Cambridge, MA: MIT Press.

Maguire, Eleanor, David G. Gadian, Ingrid S. Johnsrude, Catriona D. Good, John Ashburner, Richard S. J. Frackowiak, and Christopher D. Frith. 2000. "Navigation-Related Structural Change in the Hippocampi of Taxi Drivers." *Proceedings of the National Academy of Sciences of the United States of America* 97, no. 8: 4398–403.

Maguire, Eleanor, Katherine Woollett, and Hugo J. Spiers. 2006. "London Taxi Drivers and Bus Drivers: A Structural MRI and Neuropsychological Analysis." *Hippocampus* 16: 1091–101.

Mahon, Branford, and Alfonso Carmazza. 2008. "A Critical Look at the Embodied Cognition Hypothesis and a New Proposal for Grounding Conceptual Content." *Journal of Physiology, Paris* 102: 59–70.

Malabou, Catherine. 2008. *What Should We Do with Our Brain?* New York: Fordham University Press.

Malabou, Catherine. 2012. *The New Wounded: From Neurosis to Brain Damage*. New York: Fordham University Press.

Malafouris, Lambros. 2008. "At the Potter's Wheel: An Argument for Material

Agency." In *Material Agency: Towards a Non-Anthropocentric Approach*, edited by C. Knappett and L. Malafouris, 19–36. New York: Springer.

Males, Michael. 2009. "Does the Adolescent Brain Make Risk Taking Inevitable? A Skeptical Appraisal." *Journal of Adolescent Research* 24, no. 1: 3–20.

Mani, Anandi, Sendhil Mullainathan, Eldar Shafir, and Jiaying Zhao. 2013. "Poverty Impedes Cognitive Function." *Science* 341, no. 6149 (August 30): 976–80.

Martin, Emily. 2000. "Mind/Body Problems." *American Ethnologist* 27, no. 3: 569–90.

Martin, Emily. 2010. "Self-Making and the Brain." *Subjectivity* 3, no. 4: 366–81.

Massey, Doug. 2002. "A Brief History of Human Society: The Origin and Role of Emotion in the Social Life." *American Sociological Review* 67: 1–29.

Massumi, Brian. 2002. *Parables of the Virtual: Movement, Affect, Sensation.* Durham, NC: Duke University Press.

McCall, Leslie. 2005. "The Complexity of Intersectionality." *Signs* 3, no. 3: 1771–800.

Merzenich, Michael. 2012. "Michael Merzenich." In *The History of Neuroscience in Autobiography*, 7th ed., edited by Larry Squires, 438–75. New York: Oxford University Press.

Metzl, Jonathan. 2003. *Prozac on the Couch: Prescribing Gender in the Era of Wonderdrugs.* Durham, NC: Duke University Press.

Metzl, Jonathan. 2011. *Protest Psychosis: How Schizophrenia Became a Black Disease.* Boston: Beacon.

Moffitt, Terrie E., et al. 2011. "A Gradient of Childhood Self-Control Predicts Health, Wealth, and Public Safety." *Proceedings of the National Academy of Sciences of the United States of America* 108, no. 7: 2693–98.

Mol, Annemarie. 2002. *The Body Multiple: Ontology in Medical Practice.* Durham, NC: Duke University Press.

Moreno, Jonathan D. *Mind Wars: Brain Science and the Military in the 21st Century.* New York: Bellevue Literary Press.

Mossio, Matteo, and Dario Taraborelli. 2008. "Action-Dependent Perceptual Invariants: From Ecological to Sensorimotor Approaches." *Consciousness and Cognition* 17, no. 4: 1324–40.

Mountcastle, Vernon. 2000. "Brain Science at Century's Ebb." In *The Brain*, edited by Gerald Edleman and Jean-Pierre Changeaux, 1–36. Piscataway, NJ: Transaction.

Mozingo, Johnie, Mitzi Davis, Patricia G. Droppleman, and Amy Merideth. 2000. "'It Wasn't Working': Women's Experiences with Short-Term Breastfeeding." *American Journal of Maternal Child Nursing* 25, no. 3: 120–26.

Murphie, Andrew. 2010. "Deleuze, Guattari, and Neurosicence." In *The Force of the Virtual: Deleuze, Science and Philosophy*, edited by Peter Gaffney, 277–300. Minneapolis: University of Minnesota Press.

Nelson, Katherine, Sarah Henseler, and Daniela Plesa. 2000. "Entering a Community of Minds: 'Theory of Mind' from a Feminist Standpoint." In *Toward a Feminist Developmental Psychology*, edited by Patricia H. Miller and Ellin Kofsky Scholnick, 61–84. New York: Routledge.

Neudorf, Diane. 2004. "Extrapair Paternity in Birds: Understanding Variation among Species." *The Auk* 121, no. 2: 302–7.

Noble, K. G., M. F. Norman, and M. J. Farah. 2005. "Neurocognitive Correlates of Socioeconomic Status in Kindergarten Children." *Developmental Science* 8, no. 1: 74–87.

Noe, Alva. 2004. *Action in Perception.* Cambridge, MA: MIT Press.

Norton, Aaron, and Ozzie Zehner. 2008. "Which Half Is Mommy?: Tetragametic Chimerism and Trans-Subjectivity." *WSQ: Women's Studies Quarterly* 36, nos. 3–4: 106–25.

Ophir, Alexander G., Steven M. Phelps, Anna Bess Sorin, and Jerry O. Wolff. 2008. "Social but Not Genetic Monogamy Is Associated with Greater Breeding Success in Prairie Voles." *Animal Behaviour* 75, no. 3: 1143–54.

Ortega, Francisco. 2009. "The Cerebral Subject and the Challenge of Neurodiversity." *Biosocieties* 4: 425–45.

Oyama, Susan. 2000a. *Evolution's Eye: A Systems View of the Biology-Culture Divide.* Durham, NC: Duke University Press.

Oyama, Susan. 2000b. *The Ontogeny of Information: Developmental Systems and Evolution.* 2nd ed. Durham, NC: Duke University Press.

Oyama, Susan. 2016. "The Lure of Immateriality in Accounts of Development and Evolution." In *Mattering: Feminism, Science and Materialism,* edited by Victoria Pitts-Taylor. New York: New York University Press.

Papadopoulos, Dimitris. 2011. "The Imaginary of Plasticity: Neural Embodiment, Epigenetics and Ectomorphs." *Sociological Review* 59, no. 3: 432–56.

Papoulias, Constantina, and Felicity Callard. 2010. "Biology's Gift: Interrogating the Turn to Affect." *Body and Society* 16, no. 1: 29–56.

Pedwell, Carolyn, and Anne Whitehead. 2012. "Affecting Feminism: Questions of Feeling in Feminist Theory." *Feminist Theory* 13, no. 2: 115–29.

Petanjek, A., M. Judas, G. Simic, Rasin M. Roko, H. B. M. Uylings, and P. Rakic. 2011. "Extraordinary Neoteny of Synaptic Spines in the Human Prefrontal Cortex." *Proceedings of the National Academy of Sciences of the United States of America* 108: 13281–86.

Phelps, S. M., P. Campbell, D. J. Zheng, and A. G. Ophir. 2010. "Beating the Boojum: Comparative Approaches to the Neurobiology of Social Behavior." *Neuropharmacology* 58, no. 1: 17–28.

Piccinini, Gualtiero. 2009. "Computationalism in the Philosophy of Mind." *Philosophy Compass* 4: 512–32.

Pickersgill, Martyn, and Ira Van Keulen. 2012. "Introduction: Neuroscience, Identity and Society." In *Sociological Reflections on the Neurosciences (Advances in Medical Sociology),* edited by Martyn Pickersgill and Ira Van Keulen, xiii–xxii. Bingley, UK: Emerald Group Publishing.

Pineda, Jaime A. 2008. "Sensorimotor Cortex as a Critical Component of an 'Extended' Mirror Neuron System: Does It Solve the Development, Correspondence, and Control Problems in Mirroring?" *Behavioral and Brain Functions* 4, no. 47: 47.

Pitts, Victoria. 2003. *In the Flesh: The Cultural Politics of Body Modification*. New York: Palgrave Macmillan.

Pitts-Taylor, Victoria. 2007. *Surgery Junkies: Wellness and Pathology in Cosmetic Culture*. New Brunswick, NJ: Rutgers University Press.

Pitts-Taylor, Victoria. 2008. Introduction. In *The Cultural Encyclopedia of the Body*, Vol. I, edited by Victoria Pitts-Taylor, xvii–xxviii. Westport, CT: Greenwood Press.

Pitts-Taylor, Victoria. 2010. "The Plastic Brain: Neo-Liberalism and the Neuronal Self." *Health (London)* 14, no. 6: 635–52.

Pitts-Taylor, Victoria. 2012a. "The Neurocultures Manifesto." *Social Text/Periscope*. Accessed August 1, 2014. http://www.socialtextjournal.org/periscope/neuroculture.

Pitts-Taylor, Victoria. 2012b. "Social Brains, Embodiment and Neuro-Interactionism." In *Routledge Handbook of Body Studies*, edited by Bryan S. Turner, 171–82. New York: Routledge.

Pitts-Taylor, Victoria. 2014. "Cautionary Notes on Navigating the Neurocognitive Turn." *Sociological Forum* 29, no. 4: 995–1000.

Pitts-Taylor, Victoria. 2015. "A Feminist Carnal Sociology?: Embodiment in Feminism, Sociology and Naturalized Philosophy." *Qualitative Sociology* 38: 19–25.

Pons, T. P., P. E. Garraghty, A. K. Ommaya, J. H. Kaas, E. Taub, and M. Mishkin. 1991. "Massive Cortical Re-organization after Sensory Deafferentation in Adult Macaques." *Science* 252, no. 5014: 1857–60.

Price, Janet, and Margrit Shildrick. 1998. "Uncertain Thoughts on the Dis/abled Body." In *Vital Signs: Feminist Reconfigurations of the Biological Body*, edited by Margrit Shildrick and Janet Price, 224–49. Edinburgh: Edinburgh University Press.

Price, Janet, and Margrit Shildrick. 2002. "Bodies Together: Touch, Ethics, and Disability." In *Disability/Postmodernity: Embodying Disability Theory*, edited by Mairian Corker and Tom Shakespeare, 63–75. New York: Bloomsbury.

Prinz, Jesse. 2004. "Which Emotions Are Basic?" In *Emotion, Evolution, and Rationality*, edited by Dylan Evans and Pierre Cruse, 69–88. New York: Oxford University Press.

Prinz, Jesse. 2005. "Passionate Thoughts: The Emotional Embodiment of Moral Consciousness." In *Grounding Cognition: The Role of Perception and Action in Memory, Language and Thinking*, edited by Diane Pecher and Rolf A. Zwaan, 93–114. Cambridge: Cambridge University Press.

Prinz, Jesse. 2008. "Is Consciousness Embodied?" In *Cambridge Handbook of Situated Cognition*, edited by Phillip Robbins and Murat Aydede, 419–36. Cambridge: Cambridge University Press.

Prinz, Jesse, and Andy Clark. 2004. "Putting Concepts to Work: Some Thoughts for the Twentyfirst Century." *Mind and Language* 19, no. 1: 57–69.

Protevi, John. 2009. *Political Affect: Connecting the Social and the Somatic*. Minneapolis: University of Minnesota Press.

Protevi, John. 2013. *Life, War, Earth: Deleuze and the Sciences*. Minneapolis: University of Minnesota Press.

Puar, Jasbir. 2007. *Terrorist Assemblages: Homonationalism in Queer Times*. Durham, NC: Duke University Press.

Puar, Jasbir. 2009. "Prognosis Time: Towards a Geopolitics of Affect, Debility and Capacity." *Women and Performance* 19, no. 2: 161–72.

Puar, Jasbir. 2012. "I'd Rather Be a Cyborg than a Goddess: Becoming-Intersectional in Assemblage Theory." *PhilSOPHIA* 2, no. 1: 49–66.

Quartz, Steven. 1999. "The Constructivist Brain." *Trends in Cognitive Sciences* 3, no. 2: 48–57.

Quartz, Steven, and Terrence Sejnowski. 1997. "The Neural Basis of Cognitive Development: A Constructivist Manifesto." *Behavioral and Brain Sciences* 20, no. 4: 537–56.

Ramachandran, V. S. 2011. *The Tell-Tale Brain: A Neuroscientist's Quest for What Makes Us Human*. New York: W. W. Norton.

Raver, C. Cybele. 2012. "Low-income Children's Self-regulation in the Classroom: Scientific Inquiry for Social Change." *American Psychologist* 67 no. 8: 681–82.

Ravven, Heidi Morrison. 2003. "Spinoza's Anticipation of Affective Neuroscience." *Consciousness and Emotion* 4, no. 2: 256–90.

Reichard, Ulrich. 2012. "Conference Report: Monogamy: A Variable Relationship." *Max Planck Research* 3: 63–67.

Richardson, Sarah S. 2015. "Maternal Bodies in the Postgenomic Order: Gender and the Explanatory Landscape of Epigenetics." In *Postgenomics: Perspectives on Biology after the Genome*, edited by Sarah S. Richardson and Hallam Stevens, 210–31. Durham, NC: Duke University Press.

Rippon, Gina, Rebecca Jordan-Young, Anelis Kaiser, and Cordelia Fine. 2014. "Recommendations for Sex/Gender Neuroimaging Research: Key Principles and Implications for Research Design, Analysis and Interpretation." *Frontiers in Human Neuroscience* 8 (August 28): 650.

Rizzolatti, G., L. Fogassi, and V. Gallese. 2006. "Mirrors of the Mind." *Scientific American* 295, no. 5 (November): 54–61.

Rizzolatti, Giacomo, and Corrodo Sinigaglia. 2008. "Further Reflections on How We Interpret the Actions of Others." *Nature* 455, no. 7213: 589.

Rose, Nikolas. 2006. *The Politics of Life Itself: Medicine, Power and Subjectivity in the Twenty-first Century*. Princeton, NJ: Princeton University Press.

Rose, Nikolas, and Joelle M. Abi-Rached. 2013. *Neuro: The New Brain Sciences and the Management of the Mind*. Princeton, NJ: Princeton University Press.

Rosenblatt, Jay. 1967. "Nonhormonal Basis of Maternal Behavior in the Rat." *Science* 156, no. 3781: 1512–14.

Rostker, Bernard D., Lawrence M. Hanser, William M. Hix, Carl Jensen, Andrew R. Morral, Greg Ridgeway, and Terry L. Schell. 2008. *Evaluation of the New York City Police Department Firearm Training and Firearm-Discharge Review Process*. Santa Monica, CA: RAND Corporation.

Rotman, Brian. 2000. "Becoming Beside Oneself." Accessed August 15, 2014. http://www.stanford.edu/dept/HPS/WritingScience/etexts/RotmanBecoming.html.

Rouse, Joseph. 1998. "New Philosophies of Science in North America: Twenty Years Later." *Journal for General Philosophy of Science* 29: 71–122.

Rouse, Joseph. 2009. "Standpoint Theories Reconsidered." *Hypatia* 24, no. 4: 200–209.

Rowson, Jonathan. 2011. "Transforming Behaviour Change: Beyond Nudge and Neuromania." London: RSA Action and Research Centre. Accessed July 1, 2014. http://www.thersa.org/__data/assets/pdf_file/0006/553542/RSA-Transforming -Behaviour-Change.pdf.

Roy, Deboleena. 2004. "Feminist Theory in Science: Working Toward a Practical Transformation." *Hypatia* 19, no. 1: 255–79.

Rubin, Beatrix. 2009. "Changing Brains: The Emerging Field of Adult Neurogenesis." *Biosocieties* 4, no. 4: 407–24.

Rubin, Beatrix. 2013. "The Becoming of the Plastic Brain, a Novel Thought Style in the Neurosciences." Paper presented at the Mellon Foundation John E. Sawyer Seminar in the Comparative Study of Cultures, New York University, November 13, New York.

Rubin, Gayle. 1975. "The Traffic in Women: Notes on the Political Economy of Sex." In *Toward an Anthropology of Women*, edited by Rayna Reiter, 157–209. New York: Monthly Review Press.

Sacks, Oliver. 1995. *An Anthropologist on Mars: Seven Paradoxical Tales.* New York: Vintage.

Saldanha, Arun. 2006. "Reontologising Race: The Machinic Geography of Phenotype." *Environment and Planning D: Society and Space* 24: 9–24.

Saletan, William. 2006. "Catch and Shoot: The Perils of 'Contagious Shooting.'" *Slate.* http://www.slate.com/articles/health_and_science/human_nature/2006 /11/catch_and_shoot.html.

Samuels, Ellen. 2003. "My Body, My Closet: Invisible Disability and the Limits of Coming-Out Discourse." GLQ: *Journal of Lesbian and Gay Studies* 9, nos. 1–2: 233–43.

Saxe, Rebecca. 2009. "The Neural Evidence for Simulation Is Weaker Than I Think You Think It Is." *Philosophical Studies* 144: 447–56.

Schiebinger, Londa. 1993. *Nature's Body: Gender in the Making of Modern Science.* New Brunswick, NJ: Rutgers University Press.

Schilling, Larsen J., E. Hall, and H. Aagaard. 2008. "Shattered Expectations: When Mothers' Confidence in Breastfeeding Is Undermined—A Metasynthesis." *Scandinavian Journal of Caring Sciences* 22, no. 4: 653–61.

Schmied, Virigina, and Lesley Barclay. 1999. "Connection and Pleasure, Disruption and Distress: Women's Experience of Breastfeeding." *Journal of Human Lactation* 15, no. 4: 325–34.

Schmitz, Sigrid. 2012. "The Neurotechnological Cerebral Subject: Persistence of Implicit and Explicit Gender Norms in a Network of Change." *Neuroethics* 5, no. 3: 261–74.

Schmitz, Sigrid, and Grit Höppner. 2014. "Neurofeminism and Feminist Neurosciences: A Critical Review of Contemporary Brain Research." *Frontiers in Human Neuroscience* 8: 546.

Schwartz, Jeffrey, and Sharon Begley. 2002. *The Mind and the Brain: Neural Plasticity and the Power of Mental Force*. New York: HarperCollins.

Scully, Jackie Leach. 2008. *Disability Bioethics: Moral Bodies, Moral Difference*. New York: Rowman and Littlefield.

Sedgwick, Eve Kosofky. 2003. *Touching Feeling: Affect, Pedagogy, Performance*. Durham, NC: Duke University Press.

Shakespeare, Tom. 2013. "The Social Model of Disability." In *The Disability Studies Reader*, 4th ed., edited by Lennard Davis, 214–21. New York: Routledge.

Shepard, Gordon M. 1991. *Foundations of the Neuron Doctrine*. Oxford: Oxford University Press.

Shilling, Chris. 2003. *The Body in Social Theory*. London: Sage.

Shonkoff, Jack P., and Deborah A. Phillips, eds. 2014. *From Neurons to Neighborhoods: The Science of Early Childhood Development*. Washington, DC: National Academies Press.

Siebers, Tobin. 2001. "Disability in Theory: From Social Constructionism to the New Realism of the Body." *American Literary History* 13, no. 4: 737–54.

Siebers, Tobin. 2008. *Disability Theory*. Ann Arbor: University of Michigan Press.

Singer, Tania, and Claus Lamm. 2009. "The Social Neuroscience of Empathy." *Annals of the New York Academy of Sciences* 1156: 81–96.

Singer, Tania, Ben Seymour, John P. O'Doherty, Stephan Klaas, Raymond J. Dolan, and Chris D. Frith. 2006. "Empathic Neural Responses Are Modulated by the Perceived Fairness of Others." *Nature* 439, no. 26: 466–69.

Slaby, Jan. 2013. "Against Empathy: Critical Theory and the Social Brain." Paper presented to Summer School in Social and Cultural Psychiatry. Montreal: McGill University. Accessed August 1, 2014. https://www.academia.edu/3576043/Against_Empathy_Critical_Theory_and_the_Social_Brain.

Slaby, Jan, and Suparna Choudhury. 2012. "Proposal for a Critical Neuroscience." In *Critical Neuroscience: A Handbook of the Social and Cultural Contexts of Neurosciences*, edited by Suparna Choudhury and Jan Slaby. 29–51. West Sussex, UK: Blackwell.

Smith, Dorothy E. 1988. *The Everyday World as Problematic: A Feminist Sociology*. Boston: Northeastern University Press.

Smith, Dorothy E. 1991. *The Conceptual Practices of Power: A Feminist Sociology of Knowledge*. Boston: Northeastern University Press.

Smith, Dorothy E. 1992. "Sociology from Women's Experience: A Reaffirmation." *Sociological Theory* 10, no. 1: 88–98.

Smith, Eliot, and Gun Semin. 2007. "Situated Social Cognition." *Current Directions in Psychological Science* 16, no. 3: 132–35.

"Social Brain." 2014. London: RSA Action and Research Centre, Royal Society for the Encouragement of Arts, Manufacture and Commerce. Accessed

August 1, 2014. http://www.thersa.org/action-research-centre/learning, -cognition-and-creativity/social-brain.

Solomon, Miriam. 2007. "Situated Cognition." In *Philosophy of Psychology and Cognitive Science*, edited by Paul Thagard, 413–28. New York: Elsevier.

Southgate, Victoria, and Antonia Hamilton. 2008. "Unbroken Mirrors: Challenging a Theory of Autism." *Trends in Cognitive Sciences* 12, no. 6: 225–29.

Spanier, Bonnie. 1995. *Im/partial Science: Gender Ideology in Molecular Biology*. Bloomington: Indiana University Press.

Steinberg, Laurence. 2004. "Risk-Taking in Adolescence: What Changes, and Why?" *Annals of the New York Academy of Sciences* 1021 (June): 51–58.

Steinberg, Laurence. 2007. "Risk Taking in Adolescence: New Perspectives from Brain and Behavioral Science." *Current Directions in Psychological Science* 16, no. 2: 55–59.

Steinberg, Laurence. 2008. "A Social Neuroscience Perspective on Adolescent Risk-taking." *Developmental Review* 28, no. 1: 78–106.

Stengers, Isabelle. 2010. *Cosmopolitics I*. Minneapolis: University of Minnesota Press.

Stevens, C., B. Lauinger, and H. Neville. 2009. "Differences in the Neural Mechanisms of Selective Attention in Children from Different Socioeconomic Backgrounds: An Event-Related Brain Potential Study." *Developmental Science* 12, no. 4: 634–46.

Stueber, Karsten R. 2012. "Varieties of Empathy, Neuroscience and the Narrativist Challenge to the Contemporary Theory of Mind Debate." *Emotion Review* 4, no. 1: 55–63.

Sullivan, Shannon. 2001. *Living Across and Through Skins: Transactional Bodies, Pragmatism and Feminism*. Bloomington: Indiana University Press.

Szalavitz, Maia. 2012. "Q&A: Neuroscientist Larry Young on Sex, Drugs and Love among Voles." *Time*, November 9. Accessed October 17, 2014. http://healthland .time.com/2012/11/09/qa-neuroscientist-larry-young-on-sex-drugs-love -among-voles/#ixzz2IiNiHqrH.

Terkel, J., and J. S. Rosenblatt. 1968. "Maternal Behavior Induced by Maternal Blood Plasma Injected into Virgin Rats." *Journal of Comparative and Physiological Psychology* 65, no. 3: 479–82.

Thelan, Esther, and Linda Smith. 1994. *A Dynamic Systems Approach to the Development of Cognition and Action*. Cambridge, MA: MIT Press.

Thomson, Gill, and Fiona Dykes. 2011. "Women's Sense of Coherence Related to Their Infant Feeding Experiences." *Maternal and Child Nutrition* 7: 160–74.

Thulier, Diane, and Judith Mercer. 2009. "Variables Associated with Breastfeeding Duration." *Journal of Obstetric, Gynecologic, and Neonatal Nursing* 38, no. 3: 259–68.

Timeto, Federica. 2011. "Diffracting the Rays of Technoscience: A Situated Critique of Representation." *Poiesis and Praxis* 8, nos. 2–3: 151–67.

Tomarken, A. J., G. S. Dichter, J. Garber, and C. Simien. 2004. "Resting Frontal

Brain Activity: Linkages to Maternal Depression and Socio-economic Status among Adolescents." *Biological Psychology* 67, nos. 1–2: 77–102.

Tucker, Christine M., Ellen K. Wilson, and Ghazaleh Samandari. 2011. "Infant Feeding Experiences among Teen Mothers in North Carolina: Findings from a Mixed-Methods Study." *International Breastfeeding Journal* 6, no. 14: 14.

Turner, Bryan. 2006. *Vulnerability and Human Rights*. University Park: Penn State University Press.

Upchurch, Meg, and Simona Fojtová. 2009. "Women in the Brain: A History of Glial Cell Metaphors." *Feminist Formations* 21, no. 2 (summer): 1–20.

Valenstein, Elliot. 2005. *The War of the Soups and the Sparks: The Discovery of Neurotransmitters and the Dispute over How Nerves Communicate*. New York: Columbia University Press.

Van Anders, Sari M., James L. Goodson, and Marcy A. Kingsbury. 2013. "Beyond 'Oxytocin = Good': Neural Complexities and the Flipside of Social Bonds." *Archives of Sexual Behavior* 42: 1115–18.

Van Anders, Sari M., and Neil Watson. 2006. "Social Neuroendocrinology: Effects of Social Contexts and Behaviors on Sex Steroids in Humans." *Human Nature* 17, no. 2: 212–37.

Varela, Francesco, Evan Thompson, and Eleanor Rosch. 1991. *The Embodied Mind: Cognitive Science and Human Experience*. Cambridge, MA: MIT Press.

Vatter, Miguel. 2009. "Biopolitics: From Surplus Value to Surplus Life." *Theory and Event* 12, no. 2. doi: 10.1353/tae.0.0062.

Veenema, Alexa H. 2012. "Toward Understanding How Early-Life Social Experiences Alter Oxytocin- and Vasopressin-Regulated Social Behaviors." *Hormones and Behavior* 61, no. 3: 304–12.

Venker, Suzanne. 2013. "To Be Happy, We Must Admit Women and Men Aren't 'Equal.'" *Fox News*, February 5. Accessed March 1, 2013. http://www.foxnews .com/opinion/2013/02/05/to-be-happy-must-admit-women-and-men-arent -equal/.

Vidal, Fernando, and Francisco Ortega. 2011. "Approaching the Neurocultural Spectrum: An Introduction." In *NeuroCultures: Glimpses into an Expanding Universe*, edited by Francisco Ortega and Fernando Vidal, 7–28. New York: Peter Lang.

Vivanti, G., and S. J. Rogers. 2014. "Autism and the Mirror Neuron System: Insights from Learning and Teaching." *Philosophical Transactions of the Royal Society B: Biological Sciences* 369, no. 1644. doi: 10.1098/rstb.2013.0184.

Wacquant, Loïc. 2004. *Body and Soul: Notebooks of an Apprentice Boxer*. Oxford: Oxford University Press.

Wacquant, Loïc. 2014. "For a Sociology of Flesh and Blood." *Qualitative Sociology* 38, no. 1: 1–11.

Wahman, Jessica. 2008. "Sharing Meanings about Embodied Meaning." *Journal of Speculative Philosophy* 22, no. 3: 170–82.

Watson, Sean. 1998. "The Neurobiology of Sorcery: Deleuze and Guattari's Brain." *Body and Society* 4, no. 4: 23–45.

Weasel, Lisa. 2016. "Embodying Intersectionality: The Promise (and Peril) of Epigenetics for Feminist Science Studies." In *Mattering: Feminism, Science and Materialism*, edited by Victoria Pitts-Taylor. New York: New York University Press.

Weil, Kari. 2010. "A Report on the Animal Turn." *differences* 21, no. 2: 1–23.

Welberg, Leonie. 2010. "Mirrors, Mirrors Everywhere?" *Nature Reviews Neuroscience* 11: 374.

Wendell, Susan. 1989. "Toward a Feminist Theory of Disability." *Hypatia* 4, no. 2: 104–24.

Weston, Kath. 1991. *Families We Choose: Lesbians, Gays, Kinship*. New York: Columbia University Press.

Willey, Angela, and Sara Giordano. 2011. "Why Do Voles Fall in Love?: Sexual Dimorphism in Monogamy Research." In *Gender and the Science of Difference*, edited by Jill Fischer, 108–25. New Brunswick, NJ: Rutgers University Press.

Williams, J. H., G. D. Waiter, A. Gilchrist, D. I. Perrett, A. D. Murray, and A. Whiten. 2006. "Neural Mechanisms of Imitation and 'Mirror Neuron' Functioning in Autistic Spectrum Disorder." *Neuropsychologia* 44, no. 4: 610–21.

Williams, J. H., A. Whiten, T. Suddendorf, and D. I. Perrett. 2001. "Imitation, Mirror Neurons and Autism." *Neuroscience and Biobehavioral Reviews* 25, no. 4: 287–95.

Wilson, Elizabeth. 1998. *Neural Geographies: Feminism and the Microarchitecture of Cognition*. New York: Routledge.

Wilson, Elizabeth. 2004a. *Psychosomatic: Feminism and the Neurological Body*. Durham, NC: Duke University Press.

Wilson, Elizabeth. 2004b. "Gut Feminism." *differences* 15, no. 3: 66–94.

Wilson, Elizabeth. 2010. "Underbelly." *differences* 21, no. 1: 194–208.

Wilson, Elizabeth. 2015. *Gut Feminism*. Durham, NC: Duke University Press.

Wilson, Margaret. 2002. "Six Views of Embodied Cognition." *Psychonomic Bulletin and Review* 9, no. 4: 625–36.

Wilson, Michael. 2006. *"50 Shots Fired, and the Experts Offer a Theory." New York Times*, November 27.

Wong, Roderick. 2000. *Motivation: A Biobehavioral Approach*. Cambridge: Cambridge University Press.

Wood, Jessica, Dwayne Heitmiller, Nancy C. Andreasen, and Peg Nopoulos. 2008. "Morphology of the Ventral Frontal Cortex: Relationship to Femininity and Social Cognition." *Cerebral Cortex* 18, no. 3: 534–40.

Wood, Jessica L., Vesna Murko, and Peg Nopoulos. 2008. "Ventral Frontal Cortex in Children: Morphology, Social Cognition and Femininity/Masculinity." *Social Cognitive and Affective Neuroscience* 3, no. 2: 168–76.

Woollett, Katherine, and Eleanor A. Maguire. 2011. "Acquiring 'the Knowledge' of London's Layout Drives Structural Brain Changes." *Current Biology* 21: 2109–14.

Young, Iris Marion. 1990. *Throwing Like a Girl: And Other Essays in Feminist Philosophy and Social Theory*. Bloomington: Indiana University Press.

Young, Kimberly, Kyle L. Gobrogge, Yan Liu, and Zuoxin Wang. 2011. "The Neurobiology of Pair Bonding: Insights from a Socially Monogamous Rodent." *Frontiers in Neuroendocrinology* 32, no. 1: 53–69.

Young, Larry, and Brian Alexander. 2012. *The Chemistry between Us: Love, Sex, and the Science of Attraction*. London: Current.

Zak, Paul. 2012. *The Moral Molecule: The Source of Love and Prosperity*. New York: Dutton.

Zaki, Jamil, and Kevin Oschner. 2009. "The Need for a Cognitive Neuroscience of Naturalistic Social Cognition." *Annals of the New York Academy of Sciences*, no. 1167: 16–30.

cyborgs, 50, 52, 58, 62, 63, 142n4. *See also* extended cognition; prosthetics heteronormativity, 14–15, 52–53, 95–99, 112, 114–17, 126, 139n19, 151n11

Heyes, Cecilia, 73, 84, 146n1

Hickok, Gregory, 14, 68, 73, 146n1

hippocampus: and memory, 29–30, 138n15; and sex, 138n17

Hird, Myra, 96, 98, 115, 131n19, 133n29

horses, and behavioral synchrony, 110

Hubel, David, 25–26

hypothalamus, 31, 100, 106

Iacoboni, Marco, 68, 71, 72, 73, 80, 83, 86–87, 129n6, 134n32

Insel, Tom, 103

James, William, 24, 40, 41–42, 47, 134n1, 141n26, 145n13

Johnson, Mark, 47–49, 55, 142n3, 147n7

Jordan-Young, Rebecca, 6, 23, 32, 130n10, 132n27, 138n17, 139n19

Keen, Suzanne, 84

Kemp, Andrew, 102

Keysers, Christian, 72

kinship: affective, 96, 98, 111, 116; biological theories of, 96–100, 103–4, 111–14; feminist and queer theories of, 96–100, 115–17. See also maternal brain theory; oxytocin and attachment; oxytocin and pair bonds

Kirby, Vicki, 35, 131n19, 133n29

Kraus, Cynthia, 18, 19, 23, 140n20

Kuhn, Thomas, 120

lactation, 100, 105, 106, 107–10, 150nn8–9. See also breastfeeding

Lakoff, George, 48–49, 73–74, 83, 134n32, 142n3, 147n7

Latour, Bruno, 12, 22, 121, 133n28, 143n8

Levinas, Emmanuel, 93

Leys, Ruth, 11, 53, 68, 73, 122, 142n2

Lizardo, Omar, 9, 69, 85–86, 132nn25–26

localization thesis, 26–27, 120, 129n2

Longino, Helen, 46

long-term potentiation, 25, 136n5

Lorde, Audre, 53

Maguire, Eleanor, 29–30, 138nn15–16

Malabou, Catherine, 17, 18–20, 22, 121, 131n15, 132nn20–21, 132n26

Malafouris, Lambros, 21

Males, Michael, 27–28

Martin, Emily, 18, 56, 130n9, 131n14, 146n17

maternal body, 8. See also maternal brain theory

maternal brain theory, 99, 104–8. See also oxytocin: and attachment

memory: and dementia, 21; and emotions, 4, 45, 89; and extended cognition, 50; and learning, 17, 24, 44; and mirror neurons, 89; and oxytocin, 14, 97, 101; spatial memory, 29, 138nn14–16; working, 38

Merzenich, Michael, 29, 137–38n13

Meyer, Kasper, 89, 146n1

mirror neurons, 67–93; and aesthetic appreciation, 147–48n7; and affect theory, 11, 14; and autism, 68, 82, 86–87; and embodied simulation, 14, 69, 73, 75–76, 78–87, 92, 125–26, 134n32, 147n7; and empathy, 14; and memory, 89; and methodological critiques, 146n1

Mol, Annmarie, 12, 129n5

monkeys: in mirror neuron research, 67; in neuroplasticity experiments, 29, 137–38nn12–13

monogamy, 98, 103–4, 111–14. See also oxytocin: and pair bonds

Moreno, Jonathan, 18

Murphie, Andrew, 22, 25, 91, 136n6

Nelson, Katherine, 75, 82, 88

neomaterialism, 6, 10, 11, 60, 68, 131nn18–19, 132–33n27